The Gene Hunters

By the same author:

Norman Clark and Calestous Juma
Long-Run Economics: An Evolutionary Approach to Economic Growth (London, Frances Pinter, 1987)

The Gene Hunters

Biotechnology and the Scramble for Seeds

Calestous Juma

Princeton University Press
Princeton, New Jersey

African Centre for Technology Studies Research Series No. 1

The Gene Hunters was first published in 1989
by:
In the United Kingdom
Zed Books Ltd,
57 Caledonian Road,
London N1 9BU

In the United States of America
Princeton University Press,
41 William Street,
Princeton,
New Jersey 08540

Cover designed by Andrew Corbett.
Figures drawn by Henry Iles.
Typeset by Acorn Printing and Typesetting, Bath.
Printed and bound in the United Kingdom
at Bookcraft (Bath) Ltd, Midsomer Norton.

British Library Cataloguing in Publication Data

Juma, Calestous
The gene hunters: biotechnology and the scramble for seeds. — (African Centre for Technology Studies research series: No. 1)
1. Plants, Biotechnology
I. Title II. Series
660'.6

ISBN 0-86232-639-7
ISBN 0-86232-640-0 Pbk

Library of Congress Cataloging-in-Publication Data

Juma, Calestous.
The gene hunters.

Bibliography: p.
Includes index.
1. Plant biotechnology—Africa—Forecasting.
2. Germplasm resources, Plant—Africa—Utilization.
3. Developing countries—Economic policy. I. Title.
S494.5.B563J86 1989 338.1'6 89-3643
ISBN 0-691-04258-6
ISBN 0-691-00378-5 (pbk.)

Contents

Tables

Figures

History celebrates the battle-fields whereon we meet our death, but scorns to speak of the ploughed fields whereby we thrive; it knows the names of the king's bastards, but cannot tell the origin of wheat.

—J. H. Fabre
The Wonders of Instinct

I am not here to argue that human species ought to take responsibility for evolution on the planet, and begin through public and private institutions to make collective decisions about such matters. If that were the question . . . I would advocate that we put it off for a few centuries or more — let things run themselves while we get accustomed to the idea of evolutionary governance, develop the appropriate ethics and myths and political structures, and perhaps mature a bit. However, that is not the question before us, since we are already governing evolution.

— Walter Truett Anderson
To Govern Evolution

Acknowledgements

This study is a good example of the value of free international information exchange. Numerous researchers, industrialists, government officials and programme officers from different parts of the world have kindly shared their information with me. It is difficult to list all of them. Many of them will readily recognize their contributions to this study. I did initial work for this book while I was on a grant provided by the International Development Research Centre (IDRC) Ottawa. My discussions with Hartmut Krugmann, Steven Rosell, Chris Smart and Alyson Warhurst have helped clarify some of the theoretical and policy issues related to biotechnology. This was complemented by my association with Andrew Barnett, Norman Clark, Gerry Dempsey, Martin Greeley and Steve Pollak at Sussex. I am also grateful to Harold Miller (Mennonite Central Committee, Nairobi) and Thomas Roach (Lutheran World Relief, Nairobi) for providing me with the financial support in the early stages of preparing this book. Harold and Tom were a tremendous source of inspiration and their longstanding experience in community work helped to shape sections of this book. The preparation of this study also benefited from a grant from the Ford Foundation which enabled me to have access to the latest literature on the subject. In this regard, I would like to thank William Saint and Dianne Rocheleau (Ford Foundation) for their encouragement and support. My gratitude also goes to the Kenya Government for granting a research permit to conduct research for the Kenyan case study.

Those who generously supplied me with vital information include Adeyemi Franck Attere (International Board for Plant Genetic Resources, Nairobi), John Barton (Stanford Law School, Stanford, USA), T.T. Chang (International Rice Research Institute, Los Baños, Philippines), Sir Otto Frankel (Commonwealth Scientific and Research Organization, Canberra, Australia), Carl Göran-Hedén (Karolinska Institute, Stockholm), Goran Hyden (University of Florida, USA), Simion Imbamba (Department of Botany, University of Nairobi), Martin Kenney (Ohio State University, USA), Jack Kloppenburg, Jr. (University of Wisconsin, USA), Ward Morehouse and David Dembo (Council for International and Public Policy, New York, USA), Luis Malaret (Zoology Department, University of Nairobi), Richard Norgaard (University of California, Berkeley, USA), Robert Perdue (Agricultural Research Services, Beltsville, USA), Hope Shand (Rural Advancement Fund International, Pittsboro, USA), Hamdallah Zedan (United Nations Environment Programme, Nairobi), Susan Shen (Office of Technology Assessment, Washington, DC), Jose Esquinas-Alcázar (Commission on Plant Genetic Resources, FAO), Garrison Wilkes (Biology Department, University of Massachusetts, Boston, USA), Sheldon Krimsky (Tufts University, Medford, USA) and Norman Myers (Oxford, UK).

This study has benefited immensely from the helpful comments and suggestions of Henk Hobbelink (International Coalition for Development Action, Barcelona, Spain) who critically read the manuscript. I have also received valuable comments from Michael Chege (University of Nairobi), Hermann Field (Tufts University, Medford, USA), Robert Kates (Brown University, Providence, USA), John Kokwaro and H. N. B. Gopalan (Botany Department, University of Nairobi) and Oki Ooko-Ombaka (Public Law Institute, Nairobi). I would also like to thank Stephen Okemo (International Council for Research in Agroforestry, Nairobi) and the library staff of Sussex University, SPRU and UNEP for helping in the literature search. Their services were extremely useful in helping me locate even the most obscure publications. Finally, I would like to thank Alison Field-Juma for her support, comments and research contributions to this study.

Calestous Juma

Abbreviations

Most acronyms used in this book appear in the Appendix (institutions, conserving genetic resources). I have listed those acronyms most frequently used.

ARIPO	African Intellectual Property Organization
CATIE	Centro Agronomico Tropical de Investigacion y Enseñanza
CGIAR	Consultative Group for International Agricultural Research
CIAT	Centro Internacional del Agricultura Tropical
CIMMYT	Centro Internacional del Mejoramiento del Maiz y Trigo
CIP	Centro Internacional de la Papa
CPGR	Commission on Plant Genetic Resources
CRRI	Central Rice Research Institute
DNA	deoxyribonucleic acid
EMBRAPA	Empresa Brasileira de Pesquisa Agropecuária
FAO	Food and Agriculture Organization of the United Nations
GATT	General Agreement on Tariffs and Trade
GTZ	German Agency for Technical Cooperation
HYV	High-Yield Variety
IARC	International Agricultural Research Centre
IBPGR	International Board for Plant Genetic Resources
ICARDA	International Centre for Agricultural Research in Dry Areas
ICGEB	International Centre for Genetic Engineering and Biotechnology
ICRISAT	International Crops Research Institute for the Semi-arid Tropics
IITA	International Institute of Tropical Agriculture
IRRI	International Rice Research Institute

KSC	Kenya Seed Company
MIRCEN	Microbiological Resources Centre
NSSL	National Seed Storage Laboratory
NTBs	Non-tariff barriers
PBI	Plant Breeding Institute
PAHO	Pan American Health Organization
PBRs	Plant Breeders Rights
PPA	Plant Protection Act
PVPA	Plant Variety Protection Act
R&D	Research and Development
UNCTAD	United Nations Conference on Trade and Development
UNEP	United Nations Environment Programme
UNESCO	United Nations Educational, Scientific and Cultural Organization
UNIDO	United Nations Industrial Development Organization
UPOV	Union for the Protection of New Varieties of Plants
WHO	World Health Organization
WIPO	World Intellectual Property Organization

Glossary*

Biological diversity	The variety and variability among living organisms and the ecological complex in which they occur.
Biotechnology	The application of science and technology to the processing of materials by biological agents to provide goods and services.
Cell	The smallest unit of living matter potentially capable of self-perpetuation; an organised set of chemical reactions capable of self-reproduction. Cells contain DNA, where information is stored, ribosomes, where proteins are made, and energy conversion mechanisms.
Centre of diversity	The region where most of the major crop species were originally domesticated and developed. These centres may coincide with centres of origin.
Chromosome	In any cell, the DNA-bearing structure that carries the inheritable characteristics of the cell.
Clone	A group of genetically identical cells or organisms derived from a single ancestor.
Cultivar	An organism developed and persistent under cultivation.
DNA	Deoxyribonucleic acid. A long, chainlike, thin molecule that is usually found as two complementary chains. The chain is arranged in subunits repeated many times. The arrangement is used to store all the information necessary for life.
Embryo transfer	An animal breeding technique in which viable and healthy embryos are artificially transferred to recipient mothers for gestation and delivery.
Equilibrium	The absence of net movement one way or another.
Ex situ	Pertaining to the conservation of organisms in places where they normally do not occur. This is associated with storage facilities such as gene banks, botanical gardens, or zoos.

* Based on Drlika, 1984; OTA, 1987b; Bent et. al., 1987.

Gene	A small section of DNA which contains information for making of one protein molecule; a unit of hereditary information that can be passed from one generation to another.
Gene-pool	The collection of genes in an inter-breeding population.
Genetic diversity	The variety of genes within a particular species, variety, or breed.
Genetic engineering	The manipulation of information in the DNA of an organism in order to alter the characteristics of the organism.
Genetic resources	The useful characteristics of plants, animals and micro-organisms that are transmitted genetically.
Genome	The genetic endowment of an individual or organism.
Germplasm	A non-specific term used to refer to the genetic information of an organism or group of organisms; the total genetic variability of a species.
Hybridoma	A cell type produced by the fusion of two or more different types of cells.
In situ	Pertaining to organisms in their native environment.
In vitro	The growing of cells, tissues, or organisms in vessels under sterile conditions on artificially prepared medium.
Micro-organisms	Organisms not classified as plants or animals. These include bacteria, fungi, viruses, mycoplasma, protozoa, and cyanobacteria.
Molecule	A group of atoms tightly joined together.
Morphology	A taxonomic description of an organism, or a detailed analysis of its form and structure.
Mutation	Any change in the structure of DNA or number of genes or chromosomes in a cell.
Recombinant DNA (rDNA)	The hybrid DNA produced by joining pieces of DNA from different organisms *in vitro*.
Recombination	The breaking and rejoining of DNA strands to produce new combinations of DNA molecules.
Species	A group of closely related, morphologically similar individuals which interbreed or have the potential to interbreed; a taxonomic subdivision of genus.
Taxonomy	The study of the theory, procedure, and rules of classification of organisms according to similarities and differences between them.
Tissue culture	A technique in which portions of plant or animal are grown on an artificial culture medium.

Introduction

In his analysis of the impacts of Old World animals and plants in the Americas, Crosby described in detail the major ecological oscillations that resulted from the introduction of new life forms. He concluded: 'This wild oscillation of the balance of nature happens again whenever an area previously isolated is opened to the rest of the world. But possibly it will never be repeated in as spectacular a fashion as in the Americas in the first post-Columbian century, not unless there is, one day, an exchange of life forms between planets.'[1] These words were written in 1972. A year later, US scientists cloned the gene. Since then, scientists have marshalled enough knowledge to modify existing life forms and create new plants and animals. The world is on the verge of receiving life forms whose economic and ecological implications are likely to rival the effects of the Columbian exchange.

Writing almost at the same time, de Janvry pointed out that '[B]iological innovations have relatively little effect on labour and management requirements. They are slightly capital using and moderately yield increasing when used outside of complete packages of techniques.'[2] This assertion has neither been supported by the experiences of the Green Revolution nor advances in biotechnology. The potential impacts of advances in biotechnology will not only be irreversible, but they will also introduce major and unpredictable transformations in the global organization and distribution of production which will have far-reaching implications for Africa. This, together with its historical antecedents, is the theme of this book.

In the last decade the world has experienced major advances in technological change. The use of micro-electronics or information technology has moved from the consumer sector to the capital goods sector – the very heart of industrial production. Biotechnology, on the other hand, promises to make major inroads into the productive sector. These changes are likely to have profound effects on the world economy and the agricultural sector is likely to experience some of the earliest effects. Countries which rely on the export of high-value agricultural commodities are likely to be affected by these advances. African countries, for example, will be required to make adaptations in their agricultural practices. Their responses will depend largely on their capacity to use some of the modern techniques in biotechnology as well as conserve biological diversity as a basis for renewed agricultural production. In order to understand the advances in biotechnology and their implications, it is necessary to place the developments in a historical perspective.

Three major events related to historical botany took place exactly 200 years ago. In 1788, the ill-fated crew of the *Bounty*, under Captain William Bligh, landed on Tahiti with special instructions to move the breadfruit tree to the West Indies so as to establish a new source of food for slaves working in British colonies. The same year, the First Fleet of British colonists arrived in Australia and thus the Antipodes were turned into an extension of the European culture. In another development, British amateur scientists formed the Linnean Society in London to promote the work of the Swedish botanist, Carl Linnaeus. These three significant events were associated with Joseph Banks, a longtime president of the Royal Society and botanical adviser to royalty. It was Banks who recommended that British convicts be shipped to Australia and suggested that Linnaeus's papers and notes be purchased and brought to London. Banks also organized the mission of the *Bounty*. These events represent major landmarks in the role of genetic resources in socio-economic evolution.

In 1884-85 the European powers converged on Berlin for a conference that was to change the future of Africa. The continent was divided into numerous colonies and a long process of domination ensued. The Berlin conference was a culmination of what was later referred to as the 'scramble for Africa'. Historians have not fully explained why the scramble for colonies took place. What is clear, however, is the fact that it was partly associated with the expansion of economic activities in Europe. The scientific and technological resources available to the European powers not only contributed to local industrial and agricultural development, but it also enabled the countries to extend production to other parts of the world. These events have been used by historians to explain Africa's poor economic performance.

Most historical studies on Africa have examined the long-term changes in the control over land, labour and capital. These factors, though admittedly important, do not adequately account for the main sources of agricultural and industrial growth. Recent ecological and economic crises in Africa have led to a questioning of the validity of existing modes of analysis and the related development policies. This book aims to recast the picture by emphasizing two of the main sources of agricultural growth – plant genetic resources and the related technologies, and the accompanying institutional reorganization. The scramble for African colonies was only part of the story of European economic expansion. The other part, which is often ignored by historians, was the introduction of new genetic material and the related technological know-how into the economic system. The introduction required the reorganization of existing institutions to provide suitable conditions for the success of the proposed changes.

This book will show that colonization could have been meaningless without access to genetic resources as sources of agricultural growth. This historical legacy still dominates economic growth and has received renewed impetus from the recent advances in the capacity to modify plant and animal life. These changes, as will be shown later, pose new challenges to African economies. Africa failed in the last 100 years to benefit from the major technological advances in agricultural and industrial production. The emerging techniques in biotechnology offer both prospects and problems for the continent. While the restructuring of the global agricultural sector is likely to affect African high-value, low volume commodity exporters, the potential responses lie in the capacity to apply some of the emerging techniques to farm and industrial adjustment. The capacity to benefit from biotechnology will depend largely on the ability to conserve and utilize existing genetic resources as well as formulate science and technology policies that facilitate the process. This shift in policy focus will also require a change in the epistemological basis for the anaylsis of African events.

This study has drawn from a wide range of subjects to build a case for the conservation of genetic resources and their introduction into the economic system. The first chapter outlines the African situation and suggests an alternative approach of analysing the situation. It is suggested that the static and equilibrium notions underlying neo-classical thought are not adequate to deal with a continent that is undergoing fluctuations in a world of rapid technological change. The current conditions of technological discontinuities and socio-economic reorganization require an analytical approach that can deal with uncertainty, novelty and long term dynamics.[3] Chapter 1 traces the roots of the modern agriculture in early Western epistemology and identifies those key principles which have led to agricultural uniformity and reductionism.

In chapter 2 the reader will be introduced to the subject of historical botany. The role of genetic material in socio-economic evolution will be presented. Illustrative examples will be drawn from the use of genetic material to support the British empire. It will also be shown that the rise of the United States as a global agricultural power was as a result of the persistent efforts to introduce new genetic material and the related technology into the economic system. It is the success of this programme that has given the US the power to dominate world agricultural output.

The relatively poor performance of the Soviet agricultural system is also associated with the destruction of scientific and technological capability in plant genetics during the Stalinist era.

The US agricultural model (and the related social and political features) was later exported to a large number of Third World countries under the banner of the 'Green Revolution'. The model has taken root but has also resulted in a large number of social, economic and ecological problems. This is the subject matter of chapter 3. Chapter 4 examines the implications of recent advances in biotechnology for the Third World countries in general and African countries in particular. These innovations are likely to shift the production of high-value commodities from Third World farms to industrialized countries' laboratories. Chapter 5 deals with intellectual property issues related to genetic resources and biotechnology and their potential impact on the African countries. The new patterns of patent protection are likely to limit the flow of scientific information at a time when African countries need to build their scientific and technological capability to respond to the discontinuities analysed in Chapter 4.

The ability of African countries to take advantage of the emerging technologies depends largely on the prevailing policies on genetic resources and technology. Chapter 6 examines the history of Kenya and identifies some of the major legal and institutional obstacles to the expanded utilization of genetic resources and biotechnology. The final chapter provides policy guidelines on genetic resource conservation and biotechnology development.

This book illustrates the role played by science and technology in socio-economic evolution. It shows that economic change is largely a result of the introduction of new information or knowledge into the economic system. The manner in which that information is generated and used depends very much on the prevailing political ideology and patterns of institutional organization. African countries' capacity to undertake the required adjustments will, therefore, depend on the ability of the political leadership to grasp the significance of technological change and institutional reorganization. This will also require a shift from dependence on economic advisers and planners who still accept moribund economic theories that fail to take into consideration long-run social and economic dynamics.

The study is intended to show the options available to African countries, given their current economic and ecological crises. The continent is at the crossroads at a time when new technological and scientific advances are being made. Unlike previous technological revolutions, biotechnology offers the potential to be applied to decentralized production. It is also amenable to popular participation and can therefore be applied to the African situation. Moreover, the fact that biotechnology is science-intensive suggests that the capital-related entry barriers are minimal and the main requirement is the building of human resources through education, training and skill acquisition. But to fully comprehend the implications of biotechnology, it is important to place the subject in a long-term perspective.

Africa has found itself at a critical moment in world history. While the continent searches for alternative development strategies to deal with changes in the world commodity markets, the industrialized countries are making institutional reforms and technical advances which tend to narrow the range of development options open to the African countries. Moreover, the pace at which these changes are occurring requires African countries to make equally rapid adaptations. The future lies in the application of some of the very biotechnology innovations that are leading to the restructuring of world agriculture. Africa must innovate at technological, institutional and political levels: conventional agriculture and industry must adapt to the emerging transformations as must the intellectual as well as the policy environment.

Notes

1. Crosby, 1972, p. 113.
2. de Janvry, 1978, p. 307.
3. A full account of the proposed approach is provided in Juma, 1986 and Clark and Juma, 1987.

1. Genetic Resources and Socio-economic Evolution

US President Thomas Jefferson once said that the greatest service that could be rendered to any country was to add a useful plant to its culture. This statement underscores the importance of genetic resources in socio-economic and cultural evolution. Economic history has often focused on technological development and ignored the role of genetic material in economic change. This chapter presents a brief overview of the role of genetic resources in socio-economic change and prepares the ground for subsequent analyses of historic botany. The Jeffersonian view of the world was based on a detailed understanding of the prevailing development in the US at the time. The lessons, however, have not been adequately learnt by most African countries.

This chapter will show that the introduction of new genetic material and the related technological knowledge into the economic system is one of the most crucial sources of economic growth. To illustrate this point, however, requires an alternative epistemological basis for social analysis. Most conventional approaches are based on static notions that are inherently incapable of dealing with socio-economic systems which evolve under conditions of uncertainty. A non-equilibrium systems approach captures the destabilizing effects of new genetic material and the related technology on the socio-economic system.

Epistemology and environmental expansionism

The rise of inanimate technology and the increased use of genetic resources are closely linked. The shift from hunting and gathering to agriculture required changes in the knowledge base and the introduction of new technologies. Technology enabled mankind to introduce new modes of interacting with the environment in order to meet nutritional needs. Thus, knowledge and the prevailing material conditions were closely tied together in a non-deterministic manner. The material conditions enabled mankind to formulate a range of institutional arrangements (including traditions, myths, rituals and codes of social behaviour) which embodied some of the rules that governed mankind's interaction with the natural environment. Some of these institutional arrangements, however, acquired autonomy and became a source of instrumental power in themselves.

Knowledge of the botanical and zoological base increased the range of options available for providing nutritional needs. At this early stage of socio-cultural evolution, mankind was already showing the tendency to expand and influence the natural resource base. The decision on whether to seek control over the natural environment differed from region to region. While the American Indians, for example, adopted an ecological cosmology that avoided the need to drastically transform and control the environment, Western thought is associated with various forms of expansionism and control. One of the earliest forms of expansionism was the application of genetic resources to expand economic activities and control other human beings, a narrower domain of social behaviour that can be referred to as genetic imperialism.

Modern historians have mainly presented a truncated understanding of expansionism, often emphasizing the recent events associated with the political and economic expansion of Western Europe and the creation of colonies. Much of this has been based largely on the materialistic interpretation of the world. In order to understand the historical role played by genetic resources in social change and economic expansion, it is necessary to examine the broader philosophical basis for expanionism in a non-deterministic manner. It is understood here as the view that the role of mankind is to extend control over the rest of nature. This view is well articulated by Aristotle who conceived of a divine hierarchy over which a supreme being, God, presided and the rest of the creatures followed in a descending pecking order. Humankind, more specifically man, ruled over all the creatures below him — women, children, koala bears, snails, phytoplankton and rocks.

Aristotle flourished during the era of Alexander the Great, a significant period in Hellenic imperialism. That his philosophy should reflect such thinking is not surprising. Aristotle may have influenced his pupil, Alexander, but the extent of that influence is uncertain. Aristotelian scholasticism carried all the major elements of expansionism. He argued that the conquest of 'natural slaves' was right and therefore war against barbarians was justified. His thoughts later became central doctrines of the Judaeo-Christian tradition in which the natural environment existed mainly for the purposes of meeting human needs. Christianity became inherently expansionist in practice and philosophy.[1]

The view that humankind was supreme to all other life forms was strongly advocated by the Catholic Church. When some residents of Rome planned to organize a society to protest against the slaughter of bulls for amusement and sport, Pope Pius IX refused them permission on the grounds that animals had no souls and therefore did not deserve man's moral sympathies. Although St Francis of Assisi provides a counterpoint to the mainstream Catholic worldview, the church still remains antagonistic or indifferent to nature. With this kind of belief, all expansionists needed to do was to be convinced that other races were inferior and they could therefore justify their exploitation and even extermination.

The philosophical strand of Judaeo-Christian thought was consolidated by other notions that led to the mechanistic world in which genetic expansionism flourishes. One of the earliest advocates of expansionism was Francis Bacon. He saw the rise of science as a major source of power and tools for the control of nature. Bacon stressed that for all their pompous claims, the Greeks has not performed any experiments which led to improvements in the human condition. For him, the main goal of science was to endow human life with new discoveries and powers. He advocated the search for objective knowledge which would enable mankind to have control over all natural things.

The Baconian appeal to rationality and expansionism was furthered by René Descartes who sought to reduce all phenomena to mathematical expressions. The Cartesian world was precise and followed neat mathematical laws. With Baconian rationality one was able to identify the mathematical laws that governed the behaviour of all phenomena. The Greek view of the world as a series of chaotic events and decay was deemed irrational by Descartes and therefore dismissed as false. With the Cartesian method, the world could be reduced to separate entities which represented the whole; the behaviour of the sum of the parts was equal to the functioning of the whole. This mechanistic view of the world, which received new impetus from Galileo and Bacon, was now on its way to theoretical dominance.

Descartes compared human beings and animals to clocks composed of springs and wheels. A combination of the mechanistic view and the need to accumulate experimental knowledge led to the disregard of compassion towards animals. He defended vivisection and other cruel experiments conducted on animals in Paris and believed that the crying of animals was no more that the creaking of a wheel. Like Pope Pius IX, he saw no reason why animals deserved human compassion. Science and religion worked closely to promote expansionist attitudes.

It was Isaac Newton who consolidated the mechanistic world view in his *Principia* in 1687. He also presented a methodological synthesis of the opposing empirical, inductive method represented by Bacon and the rational, deductive method of Descartes. Newton was a strong believer in God and creation. Since God had no beginning and end, his worldview was equally timeless. What mattered for Newton was the existence of some form of equilibrium in which separate entities existed. The entities remained in balance through their mutual attraction. Gravity was the ultimate force in the Newtonian universe. It was constant and was the most fundamental of all laws. Newton's world was thus a mega-clock that worked according to some grand design — God's handicraft. The world was also reversible since it was in equilibrium and timeless. The equilibrium could be disturbed by the attraction between the constituent entities but the divine relentlessly returned to equilibrium.

Because of being timeless, Newton's worldview disregards history since events occur instantaneously and can be analysed through comparative statics. In addition, this mechanistic worldview is deterministic since all that happens is already predetermined in the initial conditions. In Newtonian systems, spontaneous emergence is severely curtailed and things change through linear progression. The Newtonian synthesis has given us a seemingly orderly world without surprises. Every entity has a predefined place in a larger constellation. And as in Aristotle, entities are defined by their fixity of form and their boundaries are clearly marked. Their positions are predetermined and do not easily change. If there are any changes, they are predictable. There is a strong causal relationship in the Newtonian worldview; for every effect there must be a discernible cause. One of the aims of research is to identify the cause-effect correspondence. In most cases, researchers look for single causes.

This mechanistic worldview has had a major influence on the development of world agriculture. Let us examine in detail the impact of Cartesian-Newtonian metaphors on botany, genetics and agricultural production. The pursuit of Aristotelian scholasticism led to a strong need to classify plants so as to understand more fully their potential contribution to human needs. And since the external form of plants was presumed to be static and discernible, an analytical method that could capture their key features could be used for classification. The belief that plants were created by God in the form they existed made it seemingly easy to classify them according to their distinctive morphological features. This mechanistic and reductionist programme was implemented by Swedish botanist Carl von Linné, popularly known by his Latinized name, Linnaeus.

The method of Linnaeus was simple, neat and appealing: one simply counted the pistils and stamens in a flower to establish its position in God's divine arrangement. With further calculation, the Cartesian part of the programme could easily be achieved and plants could be classified and neatly organized on shelves. In a few years after the system was introduced, most of Europe abandoned previous attempts to classify plants and adopted the method of Linnaeus. Like most heroes, Linnaeus succeeded in synthesizing and rationalizing existing methods of classification and analysis. His binomial nomenclature was already in use when he came to the scene. But why did Linnaeus' method prevail over the previous attempts to classify plants? The answer lies in the rise of genetic imperialism.

The clear-cut Cartesian nature of the method was extremely appealing to the scientific community. Linnaeus blossomed during the era of abstraction and reason. His method also conformed to the growing need to base botanical studies on herbarium specimens as it was becoming increasingly difficult to deal with a large body of botanical information, some of which was on plants from all over the world. It should be noted that by then imperialist countries such as Britain were already sending botanists and plant collectors to all parts of the world and the accumulated stock of specimens needed to be classified. Classification was therefore necessary if the empire was to utilize some of the plants in its expansionist designs.

In 1749 Linnaeus had written an academic thesis, *The Oeconomy of Nature*, which became a popular text in Europe and America. The book reiterated the Judaeo-Christian view that nature existed to meet the needs of mankind. His worldview was clearly Newtonian. The mechanistic and static nature of Linnaeus' work is captured by Worster: 'Essentially [the book] presents a thoroughly static portrait of the geo-biological interactions. All movement takes place in a single confined sphere, planetary in scope. Like the classical Greek naturalists, Linnaeus allows only one kind of change in the natural economic system, a cyclical pattern that keeps returning to its point of departure.'[2] According to Linnaeus, species are constantly circulating in the economy of nature with the precision of clockwork. In order to keep the system rational, God has set limits to the geographical range of each species and assigned its peculiar food. He conceived of a divine demarcation between the species and did not concern himself with variability within species.

According to Linnaeus, man must exercise his expansionist responsibility:

> All these creatures of nature, so artfully contrived, so wonderfully propagated, so providentially supported . . . seem intended by the Creator for the sake of man. Everything may be made subservient to his use; if not immediately, yet mediately, not so to that of other animals. By the help of reason man tames the fiercest animals, pursues and catches the swiftest, nay he is able to reach even those, which lye hidden in the bottom of the sea.[3]

Linnaeus' thus conformed to the Baconian ideal. With the application of science to the study of botany, mankind was able to utilize the natural endowment to fulfil his needs. As his experimental knowledge grew, so did his capacity to control and utilize nature. By doing so mankind would have realized his imperial objectives. In a rather Machiavellian fashion, the end would justify the means; the rise of human welfare, as measured by the accumulation of material wealth, would vindicate the instrumental and utilitarian ethos.

The utilitarian approach adopted by Linnaeus was also consistent with the prevailing managerial practices of British industrialists. Nature was seen largely as a storehouse for raw materials for industrial and agricultural production. This view greatly influenced the thinking of economists such as Adam Smith and led to economic theories which assumed that natural resources were inexhaustible. In addition, Newtonian equilibrium notions led to the view that pollutants and waste released into the environment would gradually dissipate as the system returned to equilibrium. The costs of ecological damage were treated as factors external to the units of agricultural and industrial production. This legacy has led to extensive environmental damage and the naive unwillingness to adopt environmentally-sound development strategies.

Linnaean thought was very popular among the Anglo-Americans. Upon his death in 1778, his books, papers and herbarium sheets were bought for £1,000 and shipped to London by James Edward Smith at the suggestion of Sir Joseph Banks, a leading advocate of the use of genetic resources for the expansion of the British empire. In 1788 amateur botanists formed the Linnaean Society to promote Linnaeus' work. The mechanistic view of botany that Linnaeus promoted faced new challenges from Darwin's *Origin of Species*, published in 1859. The theory of evolution challenged the view that species were a product of God's handiwork and did not change through time. But Darwin's concepts were easily incorporated into Linnaean thought. It was botanical business as usual.

The mechanistic and reductionist approach in the biological sciences was given renewed impetus by the discovery of hereditary factors by Gregor Mendel. Although Mendel published his work a few years after Darwin's *Origin of Species*, it went unnoticed until 1900 when it was rediscovered. The view that there were 'units of heredity', later called genes, was consistent with the reductionist Cartesian philosophy. The genes, according to Mendel, did not become diluted through progeny but were preserved; they did not change their identity but only recombined. With the rediscovery of Mendel's work, the discipline of genetics, as William Bateson called it, was born.[4] Genetics became the ultimate realization of the mechanistic and reductionist programmes in the biological sciences.[5] It was believed that every gene corresponded to specific traits in a linear way. The understanding of the functioning of the whole system was subordinated to the imperatives of genes.

Mendel's work was used to promote 'genetic determinism'. Life forms were treated as machines controlled by linear chains of cause and effect. The correspondence between genes and traits has led to simplistic interpretations of human behaviour as exemplified by socio-biology in which human behaviour is seen largely as a result of genetic make-up.[6] This deterministic interpretation of natural phenomena has been used to justify such expansionist practices as racism and sexism. Agricultural production was the first major beneficiary of genetic determinism. With Mendelian genetics, plant breeders were able to make major advances in the field. The expansionist nature of the application of genetics to agricultural production is reflected in the emergence of monocultural agricultural production. Monoculture is, indeed, genetic imperialism with all the reductionist and mechanistic underpinnings. Further advances in genetics led to new knowledge on the functioning of enzymes. This knowledge was extensively applied to industrial production, especially fermentation in the 1940s.

The most significant breakthrough in genetics research was the discovery of the physical structure of deoxyribonucleic acid (DNA) – the self-replicating molecule that carries genetic information and forms the basis of chromosomes – in 1953 by Francis Crick and James Watson. Molecular biologists have since then unravelled the basic language of life as coded by DNA. Two decades of research led to techniques that allowed scientists and industrialists to transfer specific genes from one organism to another. With these techniques, first applied in 1973, mankind is now able not only to control and use life forms, but also to modify and create new life forms. Genetic expansionism has become a pervasive phenomenon whose consequences are irreversible. This is the epistemological heritage of the biological sciences in general and agriculture in particular. The rise of biotechnology is helping push this expansionist programme further, with untold benefits and risks.

Genetic sources of social change

On 26 October 1788, the ill-fated British ship, HMS *Bounty*, under Captain William Bligh, dropped anchor in Matavai Bay, Tahiti. The ship's only mission was to collect the breadfruit tree (*Artocarpus altilis*) and introduce it into the British West Indies colonies.[7] The plant was intended to be a source of food for African slaves, or 'pieces of the Indies' as they were then called, true to the mechanistic worldview of the day.[8] The idea of sending the breadfruit tree to the West Indies was advanced by Joseph Banks, who was with Captain James Cook on his first voyage to the Pacific.[9] Banks had funded the scientific expedition, which had yielded enormous botanical and zoological knowledge on the South Pacific, including Australia.

After the voyage, Banks discussed the findings of the expedition with King George III and convinced him that the plant would be a cheap source of food for slaves in the West Indies. Then Britain had lost its American colonies and was starting to rely more on scientific activities to find alternative ways of sustaining the economy. Convinced of the potential benefits of the tree, King George III commissioned Bligh, who had sailed with Banks on Cook's second voyage on the *Resolution*, to undertake the mission. Banks made all the plans to transform the ship into a floating garden and prepare instructions for plant handling. The Royal Botanic Gardens at Kew appointed David Nelson, who had served as botanist with Cook, to accompany Captain Bligh. William Brown, another gardener from Kew, was chosen to serve as Nelson's assistant.[10]

The mission had problems right from the start. Banks and Nelson took over its running and marginalized the Admiralty. Banks provided Nelson with his own instructions, forcing the Minister of War, George Yonge, to complain that they did not expect the Captain to take orders from a gardener. Banks rejected outright the proposal that Captain Bligh be taught gardening, arguing that he had seen so much mischief done by dabblers. The entire ship was transformed into a floating garden. Even the Great Cabin, 'traditionally sacrosanct to the captain, had been taken over and enlarged, a false deck had been laid throughout the cabin and in it hundreds of round holes had been cut to receive plant pots.'[11] Botanical fanaticism ruled the place: the Great Cabin was renamed the 'Garden'.

The ship was very crowded as a result of the modifications. With a crew of 44, the gardeners could hardly find enough room to do their work. The ship sailed from Spithead on 23 December 1787. Two days later a fierce storm broke, filling the Great Cabin with water and completely ruining a large quantity of stored bread. By January 1788, the crew's allowance was reduced by 33 per cent and water was being filtered because of its bad condition. The ship was supposed to go to Tahiti by Cape Horn but the rough sea forced Captain Bligh to change course and go by South Africa arriving in October 1788. On 4 April 1789 the *Bounty* left Tahiti with 1,005 breadfruit trees for the West Indies. Crammed in an already reduced space, the crew experienced great discomfort and were also unhappy with the fact that much precious water was being given to the plants. In addition, they resented Captain Bligh's overbearing manner.

On 28 April 1789, there was a mutiny on the *Bounty* and Bligh and Nelson were put in a 33-foot open boat by the crew. The mutineers delightedly threw the plants into the sea. With their screams of 'Good riddance', they had succeeded in putting to an end the first major state-supported attempt by the scientific community to relocate economic plants. Nelson died in 1789 on Timor where *Bounty* finally landed.[12] Brown, who was kept aboard the *Bounty* because the crew needed the services of a gardener, later helped found a colony at Pitcairn. He was later shot during a land ownership dispute.

Britain did not abandon the effort to take breadfruit to the West Indies. In his second attempt, Captain Bligh watered the 1,200 plants well before leaving Tahiti. When he landed in Jamaica in 1793, the *Providence* was described as a floating garden. The breadfruit scheme, however, did not prove as successful as expected as the slaves preferred plantain and other plants.[13] It was not until the emancipation of the slaves in 1838 that the plant began to be used on a large scale; previously breadfruit was mainly eaten by pigs. Despite the fruitless efforts, Captain Bligh was honoured for his work; he was made Governor-General of New South Wales in 1805. Although he had good ideas for running a colony, his arbitrary and harsh approach led a section of the navy under Major George Johnstone to rebel. Captain Bligh was deposed and gaoled, and eventually left the colony and returned to England in 1810. Botanists have immortalized him in the genus, *Blighia*, although he is better known for the mutiny.

The breadfruit scheme demonstrates the role of genetic resources in economic development. With the help of Kew Gardens and support from the state, a major attempt was launched using the military to move a plant, whose economic value was only hypothetical, from Tahiti to the West Indies. This attempt has received little attention in history, an illustration of the failure of conventional analytical approaches to focus on the major agents of social change. Most academic traditions have paid excessive attention on the broad paths of social change and have ignored the key factors which led to social reorganization. The introduction of new genetic material into the economic system and the related technological knowledge has played a significant role in socio-economic evolution. It has taken a series of agricultural crises for this fact to be recognized. Even so, decision-makers in most African countries have not fully grasped the implications of the role of genetic resources in human welfare.

The fact that methods of moving plants have so much changed since the days of Captain Bligh has also made the role of genetic resource less obvious. Unlike mechanical technologies, which have high visibility, the genetic potential is obscured in plant material and its impact is not so obvious to the public. The contributions of genetic resources to human welfare are so widespread that they are taken for granted, and problems such as loss of biological diversity have remained unnoticed for a long time.[14] In the case of food, the range of useful resources has been narrowed to a handful. It is estimated that throughout human history over 3,000 plant species have been used for food.[15] This has been reduced to a few crops: most of the world's food comes from 20 species. Part of the reduction in the species used for food was a result of external conquest and domination which was associated with the suppression of local food crops and the introduction of exotic varieties. More recently, plant breeding programmes have reduced the number of crops used for food. In India, for example, out of 50,000 known varieties, scientific programmes focused on developing less than ten.[16]

Table 1.1
Annual production of world's major crops (1985)

Crop	Production (Million Tonnes)
Wheat	450
Corn	400
Rice	395
Potato	295
Barley	180
Sweet potato	155
Cassava	115
Soybean	105
Grapes	80
Oats	65
Sorghum	60
Sugarcane	55
Millets	50
Bananas	45
Tomato	40
Sugar beet	35
Rye	35
Oranges	30
Coconut	30
Cottonseed oil	30
Apples	30
Yam	25
Peanuts	25
Watermelon	20
Cabbage	15
Onion	15
Beans	10
Peas	10
Sunflower seed	10
Mango	10

Source: Food and Agriculture Organization, Rome.

One of the plants still suffering from prejudice is grain amaranth (*Amaranthus cruentus, A. caudatus* and *A. hypochondriacus*), which was an important food to Central and South American Indians 500 years ago, revered by both the Aztecs and Incas. When heated, the seeds burst and take on a flavour of popcorn. But, because Aztecs created idols out of popped amaranth and ate them in their religious ceremonies, the Spaniards banned its cultivation and drove it into obscurity. Although this act helped bring down their religion and culture, a few farmers continued the old tradition of growing amaranth. In the 1970s it was learnt that amaranth seeds have unusually high levels of total protein and of the nutritionally essential amino acid, lysine. This amino acid is usually deficient in plant protein, including the protein in all common varieties of major cereals such as wheat, corn and rice. Today, amaranth is being reintroduced although most cereal researchers have still never heard of it and some are cynical about its potential contribution to agriculture.[17]

The rediscovery of amaranth represents recent concern over the narrowing range of the genetic base for agricultural, industrial and pharmaceutical genetic resources. The decline of biological diversity has gradually become an issue of international concern. In the 1960s the matter seemed to interest only scientists and environmentalists. Today the subject has become a major item at various international fora, despite the limited knowledge of the extent of the problem. The 'consequences of changes in diversity cannot be forecast because our knowledge of earth's biological fabric is uneven and incomplete'.[18] The uncertainty surrounding the loss of biological diversity and its potential contribution to human welfare has led to renewed conservation efforts. Countries such as the US that have for a long time relied on the importation of genetic material for agricultural renewal have taken the lead in genetic resource collection and conservation. It is largely the understanding of the role of genetic resources and the related technological knowledge that gave the US its global lead in agricultural production.

The historical botany of the US differs remarkably from that of the Soviet Union. The Russian economy went through similar imperatives that made the need for genetic resource collection necessary. Indeed, Russian botanists, under the leadership of Nikolai Vavilov, made the hitherto most systematic global collection of economic plants. Vavilov identified eight major regions with high genetic resource diversity which he termed 'centres of origin'. He stressed that the 'enormous quantity of field and vegetable crop material discovered in these centres should be widely used in breeding work in the [Soviet Union].'[19] In order to utilize the global genetic stock, Vavilov noted, it was necessary to improve the state of Soviet climatological knowledge. Using his theory of climatic analogy, Vavilov argued that in 'selecting species and varieties of the [Soviet Union] one has to take into account the climatic conditions under which the plants introduced were growing and, wherever possible, to select varieties from regions more or less similar climatically to our country.'[20]

Vavilov's efforts were rendered fruitless by the abstractionist philosophy that dominated the Soviet Union under Stalin. With Lysenko's connivance, experimentation as a mode of scientific and

Table 1.2
Vavilov centres of crop diversity

Centre of Diversity	Crop
Mexico-Guatemala	Amaranth, avocado, beans, corn cacao, cashew, cotton, guava, papaya, red pepper, squash, sweet potato, tobacco, tomato, vanilla, tabasco pepper, guayule, luffa gourd, chayote, curuba, prickly pear, sisal hemp, sapote, sapodilla, agave, annatto
Peru-Equador-Bolivia	Beans, cacao, corn, cotton, guava, papaya, red pepper, potato, quinine, quinoa, squash, tobacco, tomato, begonia, malanga, canna, pepino, tree tomato, ground cherry, marigold, pumpkin, cocaine bush, guava, peach palm, capollin
Southern Chile	Potato, Chilean strawberry, Chile tarweed
Brazil-Paraguay	Brazil nut, cacao, cashew, cassava, para rubber, peanut, pineapple, passion fruit, mate, Surinam cherry, jaboticaba
North America	Blueberry, cranberry, Jerusalem artichoke, pecan, sunflower, black walnut, black raspberry, American red raspberry, muscadine grape, American wild gooseberry, wheatgrass, tufted hairgrass
Ethiopia-Kenya-Somalia	Banana, barley, castor bean, coffee, flax, okra, onion, sesame, sorghum, pearl millet, bread wheat, garden cress, cowpea, mustard, date palm, watermelon, canteloupe, yam, pigeon pea, Egyptian cotton, short staple cotton, tree cotton, teff, veldtgrass, wheat, lentil, lupine, safflower, coriander, fennel, khat, spurge, indigo, vernonia
Central Asia	Almond, apple, apricot, basil, broad bean, cantaloupe, carrot, chickpea, cotton, coriander, flax, grape, hazelnut, hemp, lentil, mustard, onion, pea, pear, pistachio, radish, rye, safflower, sesame, spinach, turlip, wheat, garlic
Mediterranean	Asparagus, beet, broccoli, cabbage, cauliflower, celery, broad bean, globe artichoke, sugar beet, dill, fennel, mint, desert date, rape, crimson clover, lavender, crob, chicory, hops, lettuce, oat, olive, parsnip, rhubarb, wheat, French honeysuckle, flax, carob, savory, rhubarb, fennel, thyme, rosemary, sage, hop
Indo-Burma	Amaranth, betel nut, betel pepper, chickpea, cotton, cowpea, cucumber, eggplant, hemp, jute, lemon, mango, millet, orange, black pepper, rice, sugarcane, taro, yam, balsam pear, Indian lettuce, Indian radish, tangerine, citron, lime, jambolan plum, jack fruit, bilimbi, coconut, safflower, leaf mustard, tree cotton, crotalaria, kenaf, sesbania, cardamon, indigo, madder, henna, Indian almond, senna, croton, cinnamon tree, rubber plant, bamboo, palmyra palm
Asia Minor	Alfalfa, almond, apricot, asine, barley, beet, cabbage, cantaloupe, cherry, clover, coriander, date palm, carrot, fig, flax, grape, leek, lentil, lupine, leaf mustard, oat, onion, opium poppy, parsley, pea, pear, pistachio, pomegranate, purslane, rape, rose, Russian olive, rye, sesame, wheat

Table 1.2 cont.
Vavilov centres of crop diversity

Centre of Diversity	Crop
Siam-Malaya-Java	Banana, betel palm, breadfruit, coconut, ginger, grapefruit, sugarcane, yam, mung bean, sour orange, pomelo, air potato, pokeweed, calamondin, jackfruit, durian, salacca palm, candlenut, ylang-ylang, sugar palm, cardamon, clove tree, nutmeg, curcuma
China	Adzuki bean, apricot, buckwheat, chinese cabbage, cowpea, tea, kaoliang, sorghum, millet, oat, sweet orange, peach, radish, rhubarb, soybean, sugarcane, Chinese yam, bamboo, horse radish, water chestnut, arrowhead, eggplant, cherry, hawthorn, walnut, litchi, chinaberry, sesame, Chinese cinnamon, mulberry tree, ginseng, camphor tree, lacquer tree, wax tree, aconite, ramie, hemp, fibre palm

technological enquiry was replaced by reductionist notions of dialectical materialism (and its Cartesian trappings) and Lamarckian views.[21] The triumph of Lysenko under Stalin robbed the Soviet Union of a major prerequisite for agricultural expansion – emphasis on the role of plant introduction and open experimentation. Vavilov, the man who consolidated centuries of human endeavour and brought botany into the arena of economic analysis, was arrested on a plant collection mission in the Ukraine. The aim of the expedition was to collect indigenous agricultural plants before they were displaced by new varieties imported from Russia. Vavilov was accused of agricultural sabotage and espionage. He died in prison of malnutrition of 1943.[22]

While the Soviet Union was purging its scientific establishment, especially geneticists, the US was consolidating the advances made in the field and moving into international plant breeding programmes. In 1943, the year Vavilov died, the Rockefeller Foundation launched a joint agricultural research programme with the Mexican government. This was the first time that plant breeding was used as foreign aid. These initial arrangements later contributed to diffusion of the Green Revolution. Mexico is one of the Vavilov centres. It is difficult to show a direct link between Soviet agricultural performance and Lysenkoism. It is instructive, though, that increases in food production during the Nikita Khruschev era were largely a result of increased chemical use in agriculture. These increases, however, did not help the Soviet Union reach self-sufficiency in major staples. The purging of geneticists reduced the potential contribution of modern techniques to agriculture.

Significant work such as corn hybridization, which could have increased yields by 10-30 per cent, was suppressed by Lysenko who believed that varietal inbreeding could increase crop output. The Lysenko era was marked by extreme xenophobia, and new ideas that originated from other countries were readily discredited. Thus it took 20 years for advances in hybridization to be adopted in the Soviet Union. While the US was strengthening its capacity to utilize exotic species, the Soviet Union was discouraging such work. Vavilov's efforts to search the world for varieties that could grow in the Soviet Union were used against him by Lysenkoists who accused him of wrecking Soviet agriculture. In one of his last cries of despair, Vavilov prepared a report entitled: *The Use of Foreign Agricultural Experience, the Latest Foreign Inventions, and Improved Seeds and Plants*. The report noted that there was

> a tendency to dismiss . . . all scientific work done in the capitalist world. But people forget that science and technology in capitalist countries are advanced mainly by intellectuals and people who work in science . . . Restrictions . . . imposed on the exchange of seeds have in effect halted the import of improved varieties from other countries, since that whole business can be based on a two-way exchange of innovations . . . Radical steps must be taken to correct this state of affairs.[23]

Not long after that Vavilov was arrested.

A similar fate befell Nikolai Kondratiev who argued that long cycles of economic change were associated with technological discontinuities. Kondratiev argued that the growth of economic activities was closely associated with the introduction of a cluster of innovations into the economic system.[24] With this understanding, Kondratiev had identified one of the main sources of economic growth. His ideas were later adopted by Austrian economist Joseph Schumpeter and have more recently received increased attention as a result of recent advances in technological change and its contribution to economic growth. Like Vavilov, he argued for the introduction of new knowledge or technologies into the economic system. The two arguments are analogous and focus on the main sources of economic or agricultural change. Vavilov and Kondratiev had captured the sources of socio-economic change but the smothering of their ideas by the advocates of dialectical materialism has left an indelible mark on Soviet agriculture and industry.

One of the central themes in dialectical materialism is the interaction of opposites to create a synthesis. The emphasis on opposites carries a strong Cartesian dichotomy. It is reductionist in so far as it collapses all the various features of an open system into two categories that are always in conflict. In addition, this approach assumes that change is usually linear. In open systems, the creation of new features is a complex selection process that involves both conflict and complementarity. The case for this argument is even stronger in situations that involve innovation. In such cases, successful innovations involve modifications or recombinations which rely largely on the existence of complementary conditions. This is true of the introduction of new plants as it is for technological innovations. Innovations also introduce discontinuities in social systems, a fact that weakens the case for linear progression.

Moreover, technological change tends to be non-equilibrium in so far as it keeps the economy in a state of instability and uncertainty. The requirements for state control and predictability of socio-economic behaviour seem inconsistent with the reality of evolving systems. Emphasis on uncertainty goes contrary to those aspects of Marxism that inspired the Stalinist era. The Cartesian notions in Hegel, which found their way in Marxist class analysis, were too inadequate to deal with the complex realities that characterized the rise of capitalism in Russia. It is, therefore, not surprising that both Vavilov and Kondratiev became victims of the mechanistic dialectics.

It is ironical that Marx, by adopting Darwin's metaphor, captured the essence of technological evolution. Most Marxists, however, have tended to emphasize the Newtonian aspects of Marx's work.[25] Marx considers the development of technology to be similar to that of biological organisms. 'Darwin has directed attention to the history of natural technology, i.e. the formation of organs of plants and animals which serve as the instruments of production for sustaining life. Does not the history of the productive organs of man in society, or organs that are the material basis of every particular organisation, deserve equal attention?'[26] Marx went further to apply Darwin's law of variation to the evolution of tools. He pointed out that tools evolved to adapt to particular applications and professions.

Vavilov's work and its methodological appeal carries Baconian and Newtonian notions. For Vavilov, the major agricultural crops originated from specific areas from which they spread. This enables botanists to identify the main origins of genetic resources and their trajectory of diffusion. The concept of 'centres of origin' also suggests that some specific moment in the past provided the initial conditions that form the reference point. But the fact that plants have evolved over time and space makes such designation difficult to make. But despite the criticism, Vavilov's work has indicated to plant breeders the likely sources of certain genetic material. Vavilov's centres have not only been expanded to cover wider areas, but are now considered to be associated with areas in which plant major crop domestications occurred.[27]

Vavilov's work has highlighted to the world the controversial nature of the global distribution of genetic material. The leading producers of major agricultural and industrial crops rely heavily on other countries for the supply of genetic resources. The uneven distribution of genetic resources has therefore created a certain degree of dependence among the industrialized countries on the supply of genetic resources from the Third World countries. It is difficult to estimate the degree of dependence partly because of the recombination of genetic material from various sources and the scientific and technical effort that goes into plant breeding. It is possible, however, to indicate the relative extent of the dependence of some regions on the supply of genetic resources for agricultural and industrial production.

Regions such as North America and Australia are totally dependent on external sources of genetic resources for their major agricultural crops. According to Kloppenburg and Klienman, the West Central Asiatic and Latin American regions account for 65.6 per cent of the world genetic resources for the major crops. Latin America has given the world maize, potato, cassava and sweet potato, while West Central Asia has added wheat and barley. Africa contributes some 4.0 per cent while the Mediterranean and Euro-Siberian regions add 1.4 per cent and 2.9 per cent respectively. The Chino-Japanese region contributes 12.9 per cent, Indo-China 7.5 per cent and the Hindustan region adds some 5.7 per cent.[28] The world's poorest nations as a group account for some 95.7 per cent of the world's genetic resources.

It is partly the uneven distribution of genetic resources and global food production that has led to international debate over the control of genetic resources. The fact that the so-called 'gene-poor' countries have been able to dominate world food production underscores the limitations in reducing agricultural production to the global distribution of genetic resources. In addition to these resources, success in agricultural production depends largely on the technological and scientific capability to enhance production using the available genetic resources. It is, therefore, no surprise that the industrialized countries, despite being 'gene-poor', have been able to dominate the world in agricultural production. The debate on the control of genetic resources is meaningful only if conducted in the context of broader policies and strategies for scientific and technological development.

African countries seem to be at a disadvantage on both requirements. On the basis of gross weight output Africa's genetic resources have contributed marginally to global agriculture and, therefore, have a high ratio of dependence on external sources of germplasm. Moreover, the scientific and technological knowledge of African crops is not as extensive as that of crops from other regions. Plant exploration in equatorial Africa was hampered by the devastating effect of tropical diseases such as malaria and dysentery. The history of plant hunters in Africa is a long catalogue of fateful trips. Mungo Park sent numerous plants to Britain from Pisania in 1795 while waiting to start his expedition up the Gambia. He died on his second trip. Others who died looking for plants include Christian Smith, a Norwegian botanist on Captain Tuckey's expedition up the Congo in 1816 and T. Vogel on Captain Trotter's Niger expedition of 1841.[29]

Table 1.3
Global genetic resource interdependence in food crop production[a]

	Regions of diversity										
Regions of production	Chino-Japanese	Indo-Chinese	Australian	Hindu-stanean	West Central Asiatic	Mediter-ranean	African	Euro-Siberian	Latin American	North American	Sum (%)[b]	Total depen-dence
Chino-Japanese	37.2	0.0	0.0	0.0	16.4	2.3	3.1	0.3	40.7	0.0	100	62.8
Indo-Chinese	0.9	66.8	0.0	0.0	0.0	0.0	0.2	0.0	31.9	0.0	100	33.2
Australian	1.7	0.9	0.0	0.5	82.1	0.3	2.9	7.0	4.6	0.0	100	100.0
Hindustanean	0.8	4.5	0.0	51.4	18.8	0.2	12.8	0.0	11.5	0.0	100	48.6
West Central Asiatic	4.9	3.2	0.0	3.0	69.2	0.7	1.2	0.8	17.0	0.0	100	30.8
Mediterranean	8.5	1.4	0.0	0.9	46.4	1.8	0.7	1.2	39.0	0.0	100	98.2
African	2.4	22.3	0.0	1.5	4.9	0.3	12.3	0.1	56.3	0.0	100	87.7
Euro-Siberian	0.4	0.1	0.0	0.1	51.7	2.6	0.4	9.2	35.5	0.0	100	90.8
Latin American	18.7	12.5	0.0	2.3	13.3	0.4	7.8	0.5	44.4	0.0	100	55.6
North American	15.8	0.4	0.0	0.4	36.1	0.5	3.6	2.8	40.3	0.0	100	100.0
World	12.9	7.5	0.0	5.7	30.0	1.4	4.0	2.9	35.6	0.0	100	

[a] Reading the table horizontally along rows, the figures can be interpreted as measures of the extent to which a given region of production depends upon each of the regions of diversity. The column labelled 'total dependence' shows the percentage of production for a given region of production that is accounted for by crops associated with non-indigenous regions of diversity.
[b] Because of rounding error, the figures in each row do not always sum exactly to 100.

Source: Kloppenburg and Kleinman, "The Plant Germplasm Controversy".

Table 1.4

Global genetic resource interdependence in industrial crop production[a]

Regions of production	Regions of diversity										Sum (%)[b]	Total dependence
	Chino-Japanese	Indo-Chinese	Australian	Hindustanean	West Central Asiatic	Mediterranean	African	Euro-Siberian	Latin American	North American		
Chino-Japanese	8.3	4.7	0.0	1.4	7.4	27.5	0.1	0.0	45.4	5.1	100	91.6
Indo-Chinese	5.0	43.5	0.0	7.1	2.9	0.0	22.6	0.0	18.8	0.0	100	56.4
Australian	0.0	51.2	0.0	0.0	1.8	3.3	0.0	0.0	15.4	28.3	100	100.0
Hindustanean	2.6	14.2	0.0	7.2	20.5	17.2	0.9	0.0	35.2	2.1	100	92.7
West Central Asiatic	1.5	14.7	0.0	0.0	4.5	14.2	0.1	0.0	56.6	8.4	100	95.5
Mediterranean	0.0	3.9	0.0	0.2	2.4	25.3	0.0	0.0	31.8	36.5	100	74.9
African	1.3	16.3	0.0	0.1	10.6	0.4	22.4	0.0	46.0	3.0	100	77.7
Euro-Siberian	0.4	0.0	0.0	0.1	12.8	41.3	0.0	0.0	17.5	27.9	100	100.0
Latin American	0.2	30.4	0.0	0.4	5.9	0.4	25.7	0.0	28.0	9.1	100	72.1
North American	0.0	3.7	0.0	0.0	8.3	33.1	0.0	0.0	39.6	15.3	100	84.7
World	2.1	13.7	0.0	2.0	10.8	18.2	8.3	0.0	34.4	10.5	100	

[a] Reading the table horizontally along rows, the figures can be interpreted as measures of the extent to which a given region of production depends upon each of the regions of diversity. The column labelled 'total dependence' shows the percentage of production for a given region of production that is accounted for by crops associated with non-indigenous regions of diversity.

[b] Because of rounding error, the figures in each row do not always sum exactly to 100.

Source: Kloppenburg and Kleinman, "The Plant Germplasm Controversy".

Some of the expeditions were marred by colonial rivalries. The work of Swedish botanist Adam Afzelius, which was to collect for Joseph Banks in Sierra Leone, was destroyed when the French attacked the nascent colony in 1794. The French threw his plants overboard, destroyed his plant-hutch and killed the man in charge. Despite the attack, he was able to get new supplies and continued with his collection work. John Kirk, who accompanied David Livingstone, sent a large collection of plants to Britain; one collection he sent in 1860 was mislaid in transit and recovered at Portsmouth dockyard in 1883. Later, Kirk was made the British Political Resident and subsequently Consul-General of Zanzibar from where he sent numerous East African plants to Britain. His collections included the Busy Lizzie (*Impatiens sultani*) and Pyjama Lily (*Crinum kirkii*). John Kirk was only a part-time plant collector because he had to attend to his administrative duties.

One of the most notable resident collectors was the French botanist Michel Adanson who lived on the banks of the River Senegal. Although he returned to France in 1754 with a large plant collection, his scientific contributions were minimal except for his diagnosis of the baobab tree as a member of the mallow family.[30] In 1775 the Duchess of Portland sent Henry Smeathman to Sierra Leone to collect plants 12 years before it became a British colony. After four years Smeathman had collected over 450 species. An Austrian botanist, Friedrich Welwitsch ventured beyond the confines of Sierra Leone and collected plants in Angola. His expedition was funded by the Portuguese government at the rate of £45 a month, which was less than his monthly requirements. He supplemented his income by sending out plants, insects and herbarium specimens for money. After seven-and-a-half years in the region, Welwitsch moved to London in 1863 where he lived for nine years.[31]

Most of the plants collected from equatorial Africa were of ornamental value and not economic utility. This is partly because the plants were adapted to tropical conditions different from those of Europe.[32] This also explains in part why the Royal Horticultural Society showed little interest in expeditions in this region. One notable case, however, was that of George Don who was the first botanist sent to Africa by the Society. Most of his tropical plants died on São Tomé due to cold weather. Plant collection in South Africa had a longer history of economic utility. The Dutch East India Company set up a botanical garden in 1652 at Cape Town to provide vegetables and other supplies to vessels passing by the Cape. Plant collection from Southern Africa had been undertaken long before the establishment of the gardens, but it was Paul Hermann who attempted the first systematic collection of plants in the region.[33] Subsequent notable collectors included Johan Andreas Auge, Carl Peter Thunberg and Francis Masson.

These may seem to be scientific explorations but they were closely related to the expansion of the European economic empires. Plant collection provided options for a wide range of economic utility and ornamental value. Like most human activities, the original motives were usually obscured in personal endeavour. The link between plant collection and the expansion of the empires is, however, well documented. It is part of a process by which societies cease to rely on available resources and employ scientific effort to generate new knowledge or introduce new resources into the economic system. Because of the incompatible climatic and ecological conditions, African plants did not play a major role in European expansionism.

Africa's contributions to world agriculture include sorghum, millet, palm oil and coffee. Although Africa has a wide range of genetic resources adapted to its diverse ecològical zones, the continent has not been able to develop its own genetic resources and introduce them widely into the economic system. Most of the genetic resources are still utilized at subsistence levels, and some are currently being lost with the introduction of exotic varieties for large-scale production. The use of exotic varieties has made Africa dependent for the supply as well as the importation of the related agricultural inputs and scientific and technological know-how. Expanding economic activity on the continent may in the long-run require the increased introduction of indigenous resources into the economic system in addition to the use of exotic ones.

Crisis and adaptation in Africa

Africa is currently facing a wide range of ecological and economic problems. These problems have manifested themselves in extensive ecological degradation and limited economic development. The combined effects of these problems have resulted in reduced access to basic resources such as food and energy, visibly manifested in widespread famine and declining per capita food availability. The famine is a critical stage in the weakening links between the ecological base and current forms of economic organization. While agricultural output has declined in most countries, industrial output has contributed to economic growth only marginally.

Africa was self-sufficient in food production about 20 years ago. Now the continent imports 20 per cent of its cereal requirements. Over the 1970-80 period, the quantity of cereal imported increased threefold while the cost of the imports rose by 600 per cent.[34] Africa is the only region of the world where the level of nutrition has declined in the last decade.[35] While the population grows at the pace of 3.0 per cent per year, food production expands at the rate of 1.8 per cent per year.[36] Over the 1970-84 period, the per capita production of food in Africa fell by 13 per cent.[37] About 70 per cent of Africa's population live in the rural areas where agriculture, dominated by women, is the main economic activity. The rural areas not only meet subsistence needs, but also produce food for the urban and export markets. Between 30 and 60 per cent of the GNP of African countries is derived from agriculture. The level of technology, agricultural inputs and management skills in the rural areas is still too low to provide the required food for the growing population and generate income for their increasing monetary needs. While the current capacity for agricultural production is reaching limits, increasing pressure is being exerted on the agricultural system.

These problems have created new conditions for re-evaluating Africa's development patterns and lifestyles. The crisis is viewed largely as a result of persistent droughts, an unfavourable international economic climate, and increases in external debt (and its related costs).[38] These underlying factors have contributed to chronic food shortages, worsening terms of trade between the African and the industrialized countries, and extensive ecological degradation. In response, the African countries, under the auspices of the Organisation of African Unity (OAU), have worked out a series of action plans to deal with the crisis.[39] These plans emphasize the need to strengthen the internal economic demand of the African countries. This in turn entails 'substantial expansion of capabilities at the national, subregional and regional levels for the identification, evaluation, extraction and management of natural resources and raw materials for processing to meet domestic needs.'[40] Such an expansion cannot be achieved without major changes in the scientific and technological capabilities of these countries.

The evolution of African economies in the last three decades has been marked by rapid shifts influenced by internal and external factors. The traditional economies that characterized the pre-colonial period were transformed by the introduction of new economic activities and forms of social organization. These irreversible changes have played a significant role in influencing the current situation in Africa. This, however, does not mean that the economic situation in Africa can be blamed on one primary factor. Far from it. The economic conditions are partly a result of the complex interplay between the pre-existing conditions and other factors that have emerged over the post-colonial period. The fact that the global economy is an open system makes it difficult to establish the exact causes of particular economic problems. What is noticeable, however, is a persistent lack of technological advancement, a vital aspect of economic development. Part of this can be explained by the fact that most African economies were based on raw material export and there was little or no incentive to build a strong technological base as a tool for international competition.[41] These biases were inherited at independence, and the role of technology still remains peripheral in most countries.

A few African countries are starting to recognize the role of research in economic development. Kenya, for example, has recently formed the Ministry of Research, Science and Technology and in addition has restructured the school system to reflect the long-term technical needs of the country. But much needs to be done to create a suitable atmosphere for the integration of local technological innovations into the economic system. Surveys of innovative activities on the continent suggest that there are significant innovations undertaken by the so-called informal sector which need to be recognized and supported through clearly defined policy measures. In the Congo Republic, for example, local inventors and innovators have formed a non-governmental association that enables them to promote, exhibit and exchange innovations.[42] Similar associations need to be established in other African countries.

Most African countries continued the economic structures developed over the colonial period. This, however, does not mean that they had the option to restructure and introduce technology-based economic policies. The wave of political independence came soon after the publication of the earliest quantitative studies in the industrialized countries on the role of technical change in economic development. The implications of the studies were not readily grasped by African countries, and science and technology policy continued to be ignored. The African countries have gone through two major crises which have far-reaching technological implications. The first was the 1973-74 oil price increases which led to significant technological changes in the industrialized countries. The second is the debt crisis which is closely associated with technology imports from the industrialized countries. Also related to these have been the indirect technological pressures placed on the African countries through the competition for raw materials.

Amid these changes, the industrialized countries have been undergoing a major technological revolution through the application of micro-electronics to major aspects of industrial and social organization. All along, the African economies have had to react to such changes, often without adequately understanding them.[43] Some of these changes have been undermining the very premises on which international trade is based. Concepts such as the 'comparative advantage' theory are now being challenged by evidence arising from the widespread application of revolutionary technologies such as micro-electronics, and now bio-technology, as this study will show later.[44] African countries have also continued to face the protectionist barriers placed by the industrialized countries on their goods. The industrialized countries are moving away from dependence on tariff barriers, and applying a wide range of non-tariff barriers.[45] The barriers not only undermine their potential benefits from international trade, but also raise major questions on the types of technologies and industrial adjustments these countries should introduce in order to enhance local production and compete favourably on the international market.

The 1960s were dominated by two major misconceptions about technological development in Africa. The first myth was that the existence of scientific research institutions necessarily contributed to economic change. The second was that technology was readily available and could be easily obtained through technology transfer arrangements.[46] The first misconception led to the proliferation of science and technology institutes and research councils in many African countries. These institutions were established mainly with the support of the United Nations Educational, Scientific and Cultural Organization (UNESCO). Most of them have remained marginal to the technological requirements of the countries. This should have been anticipated because scientific institutions do not emerge as abstractions but are closely linked to specific social and economic factors. Some of these institutions, however, have attempted to deal with local problems and come up with technical innovations. But most of these innovations have remained at design or prototype levels because of limited knowledge on the way technology articulates itself through society.[47]

The process of technology transfer to Africa has had its problems. The assumption that technology could easily be transferred has turned out to be misleading. The number of failed projects bears testimony to the problems surrounding the transfer of technology. Some countries have accumulated large foreign debts as a result of massive investment in large industrial projects. Some of these projects were abandoned long before they were ready for commissioning. Others are operating at low levels of capacity utilization. In addition to the genuine pre-start up problems, some African countries became victims of machinery suppliers who were more interested in making money on the supply of equipment than in the subsequent profitability of the ventures. Such patterns of technology transfer had negative effects on the recipient economies. Moreover, the high rate of technological failure has fostered an 'anti-technology' climate in some sections of the African society. Technology is seen as the main cause of some of the social problems and the view that technological failures are part of more complex socio-economic problems has received no consideration.

Some of the contractual arrangements between the suppliers of machinery and the recipients made it difficult for the African countries to build up firm-level technological and managerial capacity. In many cases the African countries did not emphasize these requirements and continued to rely on expatriate services (which had to be paid for in foreign exchange). The capacity of the African countries to negotiate for favourable technology acquisition terms was limited by inadequate information on the range of options available on the international market and their suitability to their local conditions. The rate of project failure in Africa underscores the fact that technological development is not only time-dependent but requires extensive technical, financial and managerial investment. The type and quality of investment required is not deterministic but varies with time, technology, institutional organization and industrial sector. This non-deterministic nature of technological change makes planning more difficult to undertake. The fact that most African countries undertake variants of central planning makes it even more difficult to deal with the uncertainties of technological change.

Conventional economic theory deals with the uncertainty factor by treating technology as a 'given'. Planning requires a certain measure of certainty. It also assumes a high degree of linearity, but technological change is usually non-linear and stochastic. The outcome of investment in a particular technology will also depend on the behaviour of other economic factors. One of the lessons learned from technological failure in Africa is the fact that conventional planning methods are too linear and mechanistic to deal with the evolutionary imperatives of project implementation. Planners tend to operate on the basis of targets and depend largely on the efficacy of bureaucratic rationality. No African country can claim to have fully internalized bureaucratic rationality, especially given the high level of uncertainty and lack of information on key aspects of economic and technological behaviour. Planners who operate under the assumption that there is perfect knowledge (as reflected by the single-mindedness of policy statements) have often been stunned by the rate of project failure and have at times accused economic saboteurs. African countries have yet to come to terms with the fact that project implementation is a learning process in which much knowledge needs to be generated on site.[48]

The short-term nature of most development plans limits the scope for incorporating long-term technological considerations. The uncertainty and diversity often associated with technological change requires flexible policies and institutional arrangements for their implementation. So far, very little is known about the technological behaviour of African countries; there are no historical precedents to go by. Past developments elsewhere can only give possible indicators but cannot serve as an adequate source of policy insights because of the significant historical differences between those experiences and the prevailing conditions in Africa. The African technological situation should be viewed against this complex historical background.

Divergent interpretations

Africa has recently become the researcher's nightmare; theoretical approaches that have hitherto been upheld as suitable for analysing the

African situation have consistently failed, and academics have been forced to resort to technocratic methods which rely on the barren presentation of statistics. The approach of reducing socio-economic complexity to technical problems is still popular among planners and development agencies. But in some research fields, for example anthropology, there have been attempts to re-examine the conventional view of Africa in light of the prevailing reality. Economists, on the other hand, have made very marginal adjustments in their view of the African situation and as a result no genuine understanding of the situation seems to be emerging from the plethora of studies on Africa being released. The picture that is normally presented is that of a dying continent. Analysis is giving way to obituaries.

Nearly all the statements made about the future of Africa are based on linear projection of current trends. A recent summary of the conventional interpretations of the African situation shows that: 'In 2057, Africa is a continent of 2 billion people, one-fifth of the world's total population, with an average per capita income of $3,800 and an average life expectancy of 76, with almost all its children in primary school and half in secondary, and with intensive use of its rich natural resources. . . . As such, it is about as densely populated, as wealthy, healthy, and educated, and as environmentally transformed as Greece was in the early 1980s.'[49]

Table 1.5
Estimated and projected trends in Africa

	Africa			Greece
SELECTED INDICATORS	**1957**	**1987**	**2057**	**1980–85**
Demographic Variables				
Total population (millions)	263.0	599.0	2,154.0	—
Population growth (%/year)	2.3	3.1	0.5	—
Infant mortality (per 1,000)	180.0	101.0	12.0	16.0
Life expectancy at birth (years)	40.0	53.0	76.0	74.0
Level of urbanization	18.0	31.0	75.0	59.0
Economic Variables				
GDP per capita (1980$)	44.0	815.0	3,800.0	3,790.0
Capital goods production (million 1975 US$)	133.0	1,273.0	75,000.0	—
Food supply per capita (calories)	2,095.0	2,294.0	3,353.0	3,627.0
Human Resource Variables				
Literacy (%)	15.0	53.0	80.0	92.3
Scientists and engineers in R&D (per million inhabitants)	13.0	103.0	266.0	279.0
Natural Resources and Environmental Variables				
Arable land (hectares/person)	0.7	0.4	0.15	0.4
Energy consumption per capita (kg coal equivalent/person)	172.0	541.0	2,069.0	2,062.0

Source: Alan Shawn Feinstein World Hunger Program, *Projecting Current Perspective.*

Agriculture is still the main source of economic activity in Africa and is expected to play a significant role in social change. Conventional economic analyses project that the sector is not expected to experience any significant renewal in the next 30 years. The expected growth rates, according to the Food and Agriculture Organization (FAO), will not be enough to offset the Malthusian spell that is cast upon the continent. FAO predicts that food production will stagnate in West, Central, and Sahelian Africa and will decline in East Africa by the year 2000; increases will be recorded in Northern and Southern Africa.[50] The main sources of increased food production are expected to be higher yields, cultivation expansion and land-use intensification. But these will not be at levels significant enough to cope with food shortages. Imports will, therefore, be needed to supplement local production. The Economic Commission for Africa (ECA) estimates that by 2008, the import bill to meet cereal shortfalls in Africa will be US$14 billion (at 1980 prices).[51]

Africa's role in international trade has been mainly that of supplying raw materials, especially agricultural products. This role has a specific place in conventional economic theory. According to the principle of comparative advantage, these countries should benefit from their participation in international trade. This principle rests on the notion of relative production costs: the relative efficiency of producing one product in comparison to others in the same country; compared with the relative efficiency in other countries. Ricardo thus held that trade between countries in products they both could produce at relatively lower prices would be beneficial to both countries.

This model, which is based on equilibrium notions developed by Newton, has been used to argue for that participation of the African countries in international trade as raw material producers. These countries have introduced policies aimed at attracting foreign investment and extractive technology for the mining and agricultural sectors. But recent developments in the industrial and agricultural sectors show that the main agent of long-term economic development is not necessarily the relative endowment of land, labour and capital (factors of production in neo-classical parlance) but the capacity to undertake technological innovation. It is science and technology that largely defines the global distribution of productive activity. It can be argued that African development has lagged behind not necessarily because it has been exploited, but because the exploitation has gone hand-in-hand with limited scientific and technological development.

In order to provide a more realistic picture of Africa in relation to the changes in the global technological map, it is important to conceive of social systems as non-linear. They change through the introduction of new information and technologies as well as institutional reorganization. For this reason the introduction of technological systems into the economy reorganizes it and allows it to change its methods of production and patterns of resource utilization. New information may also be brought into the economic system through the introduction of new plants. The process of plant domestication was one of the earliest ways of transforming the economic system and allowing it to realize complexity. With this act, mankind was able to move from one level of production to another. And with the introduction of new genetic material also came a set of technologies and information pertaining to the management of production.

The introduction of new genetic material and the related technological knowledge into the economic system leads to discontinuities in production methods and reorganizes the prevailing social relations. The reverse may also happen: the reorganization of social relations may make possible the introduction of new genetic material or technological knowledge. But organizing social relations alone without the capacity to generate technological options or provide a range of genetic material is futile. Social change is therefore full of surprises; emphasis on prediction often leads to a false sense of confidence. Seemingly minor introductions of technology or genetic material into the economy may cause major changes in the system.

The uncertainty factor in social systems makes the process of change non-deterministic. This view would suggest an alternative approach to economic planning. In the first place, a situation like that in Africa, which is characterized by major ecological and economic crises and is undergoing major fluctuations, can undergo major transformations in very short periods. Such a system lends itself to experimentation. The potential for adopting an inappropriate development path, however, is also high. Economic experiments under such conditions need to be undertaken on a small scale, a factor that underscores the prevalence of community groups in Africa and their growing politicization. It is important to emphasize that since economic change is irreversible and long-term by nature, the choice of development options needs to start on a small scale.

On the whole, the role of genetic resources in socio-economic evolution is often taken for granted. In order to fully comprehend the role of genetic resources and the related technological knowledge, it is necessary to adopt an analytical approach that can deal with discontinuity and change through time. Such an approach should present a historical view of the changing role of genetic resources in socio-economic evolution. The next chapter presents an exploration into historical botany — an attempt to show the complex linkages between social change and genetic resources through time. The exploration will provide a valuable basis upon which to analyse further the linkages between genetic resources, technological change and socio-economic evolution.

Notes

1. 'By destroying paganism, Christianity made it possible to exploit nature in a mood of indifference to the feelings of the natural objects. It is often said that for animism the Church substituted the cult of saints. True; but the cult of saints is functionally different from animism. The saint is not *in* natural objects; he may have special shrines, but his citizenship is in heaven. Moreover, a saint is entirely a man; he cannot be approached in human terms. In addition to saints, Christianity of course has angels and demons inherited from Judaism and perhaps, at one removal, from Zoroastrianism. But these were all as mobile as the saints themselves. The spirits *in* natural objects, which formerly had protected nature from man, evaporated. Man's effective monopoly on spirit in this world was confirmed, and the old inhibitions on the exploitation of nature crumbled'. White, 1967, p. 1205.

2. Worster, 1985, p. 34.

3. Quoted ibid., p. 36.

4. William Bateson was a fervent advocate of Mendel's work. He named his youngest son, Gregory after Mendel. Ironically, Gregory Bateson was later to become a strong advocate of the systems approach to natural phenomena. See, for example, Bateson, 1979.

5. See Capra, 1982, for a review of the impact of Newton on the sciences.

6. See Wilson, 1975, for the principles of sociobiology.

7. 'Commonly, the breadfruit is prepared like the potato or the plantain. Fired, it tastes like French fried potatoes; boiled, it is nearly tasteless, unless eaten with sauces. [The] preference is for baked breadfruit. The whole fruit is put in the coals of a fire and baked until the outer skin is charred. The skin is then peeled and the hot pulp eaten,' Oster and Oster, 1985, p. 37.

8. 'The English efficiency experts reasoned that breadfruit would be the ideal fuel for the slave machine,' ibid, p. 35.

9. It was Banks who suggested to the British government that criminals be sent to Australia. The First Fleet arrived in Australia in 1788, the same year that the *Bounty* landed on Tahiti and the Linnaean Society was formed.

10. Lemmon, 1968, p. 95.

11. Ibid., p. 95.

12. The genus, *Nelsonia*, of the acanthus family, was named after him.

13. Although the breadfruit experiment failed, Captain Bligh's collection of the thick cane from the Pacific was widely grown in the West Indies.

14. See Myers, 1979 and 1983.

15. This is possibly a low estimate given the limited extent of knowledge on the temporal and spatial distribution of genetic diversity and the changing patterns of human utilization and conservation.

16. Alvares, 1987, p. 11.

17. Vietmeyer, 1986, p. 1379. Amaranth is now being bred in the US for large-scale production. The US Agency for International Development (AID) recently gave a US$750,000 grant to the US National Academy of Sciences to study the improvement and adaptation of amaranth in Guatemala, India, Kenya, Mexico, Peru and Thailand. See Tucker, 1986, p. 13, and National Academy of Sciences, *Amaranth*.

18. Wolf, 1987, p. 6.

19. Vavilov, 1951, p. 45.

20. Ibid.

21. Lamarck believed that characteristics acquired through exposure to the environment could be passed on to subsequent generations. Lysenko followed this logic and argued that plants that were physically modified could pass on their characteristics to their progeny thereby eliminating the need for conventional breeding. This view as also consistent with the ruling ideology. New Soviet ideology could be passed on to subsequent generations thereby eliminating the fear of inherent characteristics or dominant bourgeois features.

22. The death of Vavilov was part of a protracted purge that saw the liquidation of leading Soviet geneticists. For details, see Zirkle, 1949a. Additional discussion on Lysenkoism is found in Joravsky, 1970, Medvedev 1969, and Levins and Lewontin, 1985.

23. Quoted in Popovsky, 1984, pp. 112-13. Lysenko's reign ended with the growing realization that Soviet Agriculture could benefit from advances made in other countries. It took Khruschev to personally visit the Garst farm in Iowa to observe the success of US breeding programmes. The Soviet Union started importing hybrid seed in 1955. 'It is indeed an irony of fate that Garst's business was a branch of a firm founded by the plant breeder Henry Wallace, who had been the [US] secretary of agriculture to whose farming experience Vavilov had tried in vain to direct attention back in 1938,' Popovsky, 1984, p. 103.

24. For a collection of studies on the theory of long waves, see Freeman, (ed.) 1983.

25. This theme is extensively analysed in Juma, 1986, Chapter 1 and Clark and Juma, 1987.

26. Marx, *Capital*, Vol. 1, p. 493. Marx found out that he could not study capital accumulation without having a detailed understanding of technology. All his work on technology has yet to be published. Marx's recently published works on technology are contained in Müller, 1981; and Winkelmann, 1981.

27. See Zhukovsky, 1975.

28. Kloppenburg and Klienman, 1987c, p. 194. These figures are based on weight and not market value and serve only as indications of the global interdependence on genetic resources.

29. One way of re-reading European economic history is to study the triumph of military expeditions; the other is to study those who combed the

world for economic plants.

30. It was thus named *Adansonia digitata*.

31. His name is immortalized in the *Welwitschia bainessi*, an extraordinary xerophytic plant of a very large size.

32. Attempts to collect plants in the Ugandan highlands by the British Museum in 1906 and 1934-35 did not yield plants of economic significance. Well-known horticulturalists – George Taylor (later Director of Kew Gardens) and Patrick Synge (prominent member of the Royal Horticultural Society) – comprised the later expedition.

33. Carl Linnaeus said in his *Flora Zeylanica*: 'Good Lord! How numerous, how rare and how wonderful were the plants that presented themselves to Hermann's eyes!'

34. For details, see Christensen and Witucki, 1982. The 'region's cereal import bill climbed from $600 million in 1972 to $4.5 billion in 1983, a ninefold increase. By 1984 food imports claimed some 20 per cent of total export earnings'. Brown and Wolf 1985, p. 8.

35. OTA, 1984b.

36. World Bank, 1984.

37. UN, 1986a.

38. See UNDP, 1986. By the end of 1985, Africa's external debt was estimated at US$150 billion. Of this, 40 per cent was owed by four North African countries. The share of Sub-Saharan Africa was US$90 billion. The debt share of the low income countries was US$30 million and represented 55 per cent of GDP. The share for the whole of Africa was 35 per cent.

39. The most important of which is the 1979 Lagos Plan of Action and the 1980 Final Act of Lagos.

40. UNDP, 1986, p. 4.

41. The use of technological change as a competitive tool is analysed by Abernathy et al., 1983.

42. Dianzungu, 1987.

43. Some of the trade and development impacts of the emerging technologies such as micro-electronics are outlined in UNCTAD, 1986a.

44. See Kaplinsky, 1982, for a critique of the comparative advantage theory.

45. There are hundreds of non-tariff barriers (NTBs) to international trade. NTBs are defined as all those public regulations and governmental practices which introduce unequal treatment between domestic and foreign goods of the same or similar production. They fall into five major categories: quantitative restrictions; non-tariff charges; government participation in trade; customs procedures and administrative practices; and technical barriers. The concept of comparative advantage loses its meaning under conditions of trade protectionism. For details on NTBs, see Deardoff and Stern, 1985, pp. 13-14.

46. The 1962 UN conference on science and technology for the undeveloped

countries held in Geneva asserted that the developing countries could 'leap across the centuries' by applying the technologies that were already available on the international market and did not need to go through their own industrial revolutions by reinventing some of the available technologies. The conference also stressed that there were no vested interests that could have hindered the transfer of the technologies to the underdeveloped areas, United Nations, 1963, p. 3. The failure of the high hopes led to the 1979 UN Conference on Science and Technology for Development (UNCSTD) held in Vienna.

47. Recent attempts to introduce seemingly simple innovations, such as improved stoves, in Africa have revealed how social complexity shapes technology and vice versa.

48. This theme is explored in detail in Juma, 1988.

49. Alan Shawn Feinstein World Hunger Program, 1987, p. 6. It is not clear whether the statement is a commentary on Africa then or Greece now.

50. FAO, 1986b, p. 9.

51. ECA, 1983, p. 29.

2. Explorations in Historical Botany

The growth of agricultural complexity in Europe became more salient in the 16th and 17th centuries. This period is also marked by dramatic increases in the pace of botanic exploration and the identification of agricultural and industrial plants. This chapter traces historical and economic botany from the earliest known records and covers the early plant hunters, the role of genetic resources in the expansion of the British empire and the rise of the US as an agricultural superpower. Much of what the African countries have adopted as modern agriculture is rooted in the way genetic resources were incorporated in the socio-cultural evolution of the Western countries. Historical and economic botany show the links between economic complexity and the introduction of new genetic resources (as well as the related knowledge and technology) in agricultural production.

Early plant hunters

Seed collection is as old as human civilization. Some of the most interesting studies of socio-cultural evolution relate to human efforts to domesticate plants.[1] Seed collection and the introduction of new crops are part of human history though they are often taken for granted. Seed collection and plant introduction were not reserved for any particular individuals, but records of early plant introductions are closely associated with state authorities. This does not mean that individual efforts never played a major role in plant introduction, far from it. Plant collection expeditions supported by the authorities receive greater attention and are often recorded. Indeed, most historical records emphasize the role of heroes and ignore the evolutionary imperatives that influence their decisions. The case of plant collection is even more personalized. There is another way of looking at the role of state authorities in plant collection: such expeditions were usually risky and uncertain; they required support and authorities that could underwrite their costs. The fact that the authorities were willing to commit resources to plant collection illustrates their economic and political value to the nobility and the country.

One of the earliest recorded plant collection expeditions took place in 1495 BC when Queen Hatshepsut of Egypt sent a team under Prince Nehasi ('The Negro') to the Land of Punt (Somalia/Ethiopia) to obtain specimens of the plants whose fragrant resins produced frankincense, a species of *Boswellia.*[2] The expedition brought back 31 young trees which were planted in the garden of the Temple of Amon at Thebes. This expedition had economic motives and focused on a particular plant, although ebony, gold, ivory and other products were also brought back to Egypt. Inspired by the expedition, the Queen's nephew and successor, Thutmosis III, returned from a military expedition in Syria with a large number of plants of no obvious economic value.[3] The Egyptian civilization was partly based on the introduction of plants from the neighbouring regions. Plant collection was part of the tradition and new plants were often brought home after military expeditions.[4]

Similar expeditions were recorded in other parts of the world. An inscription in Mesopotamia tells of Sargon crossing the Taurus Mountains to Asia Minor collecting trees, figs (*Ficus* species), vines and roses for introduction in his country about 2500 BC. The Japanese erected a monument at Kamo-Mura in Wakayama province to honour Taji Mamori who was sent by the imperial authority to China in 61 AD to collect citrus fruits.[5] These expeditions were often based on the understanding of the close links between genetic resources and economic growth. This was clearly articulated by the Spanish royalty: when, in the 18th century, Hipolito Ruiz and Antonio Pavón were sent to collect plants in Peru and Chile, the Spanish crown made it clear that scientific enquiry, botanic gardens and economic interests were closely linked.[6] Ruiz, Pavón and the Frenchman Joseph Dombey spent ten years collecting plants in Chile and Peru. Although their work was mainly taxonomical, they were equally interested in the economic value of the plants they encountered. They took keen interest in the medicinal properties of quinine, for example.[7]

The bewildering range of plant species could not be handled without some form of organization or classification. Various cultures evolved their own ways of classifying plants.[8] The systematic study of plants in Western culture took root in Ancient Greece, although a papyrus record from Egypt dated from 1600 BC contains a list of medicinal plants and their uses. Similar lists from the Assyrians date from 700 BC. Much of the early Greek work on plants was associated with medicine. Hippocrates and other Greeks made their contributions to the subject over the 460-370 BC period.[9] It was Aristotle and Theophrastus (students of Plato) who undertook what can be called botanic studies.

Theophrastus (370-287 BC) took over from Aristotle as head of the Lyceum in Athens to which he bequeathed his garden with 450 species; this is the earliest botanic garden on record. Theophrastus wrote two books in which he made the distinction between monocotyledonous and dicotyledonous plants. His work extended to a study of plant ecology.

Alexander the Great wanted his former tutor, Aristotle, to write a book on natural history and so during his Asiatic campaign of 331-323 BC, he employed special people to hunt, fish, hawk and collect plants. One of the plants he saw during the campaign was the banana (*Musa* species), which originated in the Indo-Malayan region and was already being grown widely in India. He thought the fruit was unwholesome and prohibited his soldiers from eating it. Aristotle never wrote the book but the papers were acquired by Theophrastus who used them to prepare his *Enquiry into Plants.*[10] It was during this period that the aromatic basil (*Ocimum basilicum*) and the everlasting amaranth (*Gomphrenia globosa*) were, presumably, introduced into Greece. Most botanic studies after this period were associated with herbal medicine, as communities or individuals searched for the therapeutic properties of plants, and as a process of cumulative socio-cultural evolution, selected, retained and evaluated the carriers of such useful properties.

The movement of plants in Europe was closely associated with political developments. The Romans spread the plants during their military expeditions. Some plants had been moved from Egypt to Greece and finally into the Roman empire as political power changed hands. 'Indirectly the finds of those earliest plant hunters of the Nile, of the Biblical lands and the Mediterranean seaboard, were now to become part of the British landscape so far as the climate would allow. For the Romans, as they settled in their forts and their villas up and down the length of the country, planted their native vegetation.'[11]

The movement of plants was later associated with the spread of Christianity, especially through monasteries. The period of the Crusades helped spread the plants to different parts of the world as well as bring new species back to Britain. But the challenge was enormous, as effective plant transportation techniques had not been developed. The exchange of seed continued after the Crusades as Britain became more integrated with Europe through trade and diplomacy; seeds and plants became part of the diplomatic bag sent from the Continent to King Henry VIII. We are told that he developed 'a desire for the unusual and rare so that seeds and plants would often accompany state papers in the ornate wallets of courtier and statesmen diplomats of the day on their travels throughout Europe.'[12]

The expansion of economic activities in Europe in the Middle Ages was associated with the increased use of plant products, some of which, such as pepper, could not be easily obtained locally. By the 16th century,

Venice had become a rich and beautiful city partly because of profits from the pepper trade. When the Turks blocked the overland trade routes east of the Mediterranean around 1470, Portuguese, Italian and Spanish explorers set out to find alternative sea routes to the east; this led to the discovery of the Americas.[13] In his extensive account of plant history, Hobhouse says: 'So important — but at the same time so pedestrian — was pepper a cause, that in the five hundred years since the Turks first created the problem historians have invented all sorts of other reasons for the European voyages which pepper inspired. What relatively new nation, such as America, wants to be offered the prosaic motives for its discovery? What child wants to know that it is merely the outcome of an apparently casual affair?'[14] This, however, ignores the fact that spice trade was far from a casual affair, it was part of complex economic developments in Europe.

Spices had an important economic function: they helped to introduce variety in basic foods, and would have been difficult to replace. New combinations of tastes could be introduced in the diet by innovative cooking with spices.[15] In addition, they were used as preservatives and also to conceal the unpleasant flavour of meat or fish that was not fresh. This was an important function since there were no effective ways to keep such foods over a long period. This was also the period of rapid urban growth in Europe, and small trading nodes became major centres of economic transaction with a growing demand for eating houses.[16] What started as a search for spices led to major changes in world history. Colonization was one outcome of these economic changes and exploration. The acquisition of colonies made land and labour available to the colonizers, and colonies could be transformed into surplus-producing economies only through the expanded cultivation of existing crops and the introduction of new ones. Most historical assessment of economic development has focused on labour, land and capital but ignored the basic source of agricultural output — plants. The increased application of plants to economic growth was accompanied by institutional and technological innovations in the field of plant transportation.

Institutional and technological innovation

One significant category of institutional organization related to plant collection is the botanic garden.[17] As already indicated, much of the plant knowledge accumulated was related to medicine. In this respect, it was necessary to identify and make available the main sources of medicinal properties. Surgeons could not operate without ready access to, and relative control over sources of medicine — plants. Not surprisingly,

Physick gardens emerged all over Europe, especially in the 16th and 17th centuries. The fact that most of these gardens were attached to university medical faculties shows the close links between genetic resources and knowledge. By 1545 there were already botanic gardens in Padua, Florence and Pisa and others were established: at Leiden in 1577, Leipzig in 1579 and Montepellier and Heidelberg in 1593. Louis XIII's personal physicians were responsible for the formation of the Jardin des Plantes in Paris in 1635 for the purpose of teaching medical students. A Physick garden initiated at Edinburgh in 1690, later became the Royal Botanic Garden of Edinburgh.[18] In addition to their home botanic gardens, the European powers began replicating these in their colonies: the Dutch at Cape Town in 1694; the French on Mauritius in 1733; and over the 18th century the British on St. Vincent and Jamaica and in Calcutta and Penang.

Kew Gardens began as an informal royal garden in 1759 and became a public botanic garden in 1841. The institution later played a key role in the transfer of plants such as oranges, bananas, pineapples, mangosteen, almonds, tung oil seeds, cochineal cactus, chaulmoogra, ipecacuanha, pyrethrum and mahogany to other parts of the world. The Amsterdam and Paris botanical gardens were also active in the transfer of plants to the colonies. The global production of coffee, for example, was facilitated by these gardens.

Coffee (*Coffea arabica*) originated in the highlands of Ethiopia and was first domesticated in the Arab world; from there it presumably reached India. The Dutch found the plant in India and planted it in Ceylon in 1659 and Java in 1696. A single coffee tree which reached the Amsterdam Botanic Garden in 1706 was later the basis of most of the coffee grown in South America. Seeds from the plant were first sent to Surinam in 1715, and coffee trees from Surinam were transferred to Brazil in 1727. Seeds from the same tree in Amsterdam were sent to the Jardin Royal in Paris and subsequently to Martinique in 1723.[19]

The tradition of coffee consumption was soon established in the colonies and production expanded. Coffee production in African countries began much later despite the fact that Africa was the origin of the plant. Kenya, one of Africa's principal coffee-growers, began production in the 1890s. It was John Patterson who brought the seeds from Aden (now in South Yemen) and planted them at Kibwezi near the coast in 1894. The climate and soil there were unsuitable and the production of seedlings for up-country plantations was later taken up by Catholic missionaries in Nairobi. Coffee in Kenya was produced mainly on large farms; currently its production is regulated by the government.[20]

A series of events in Europe in general, and Britain in particular, led to significant institutional changes which later affected the pace of plant

Fig. 2.1 :

Botanical gardens used to spread tropical crops.

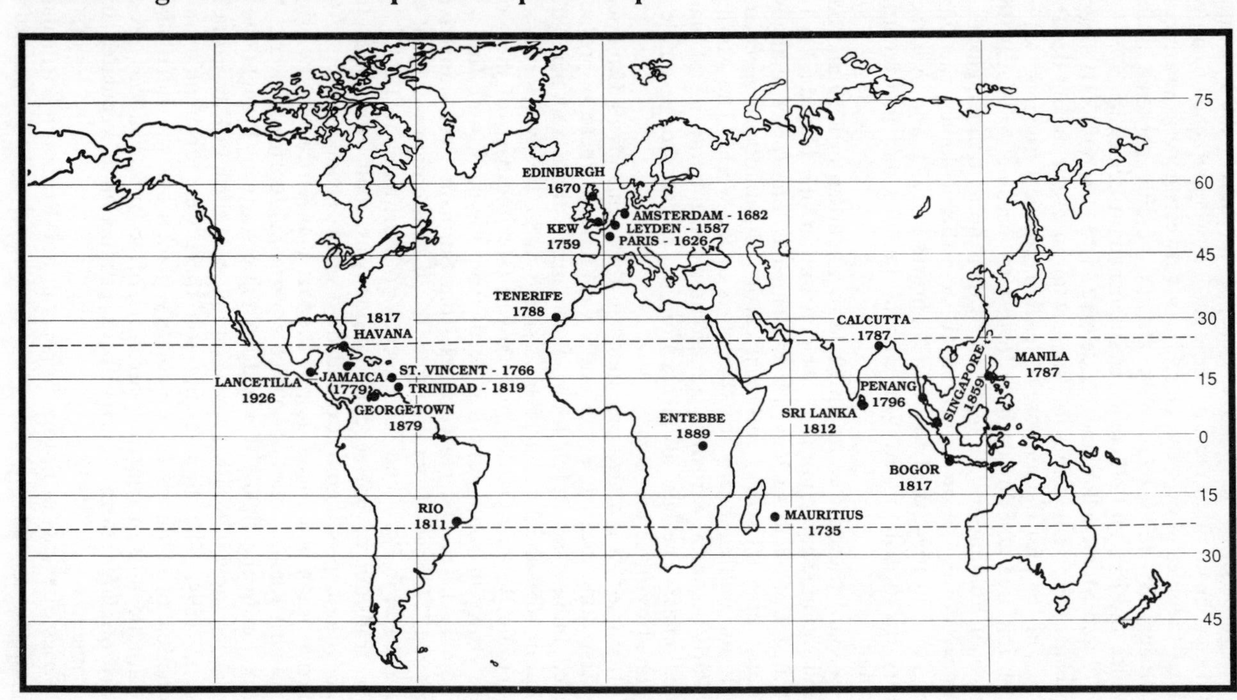

Source : Plucknett et al., Gene Banks, p. 46.

Fig. 2.2 :

Distribution of the cultivated coffees.

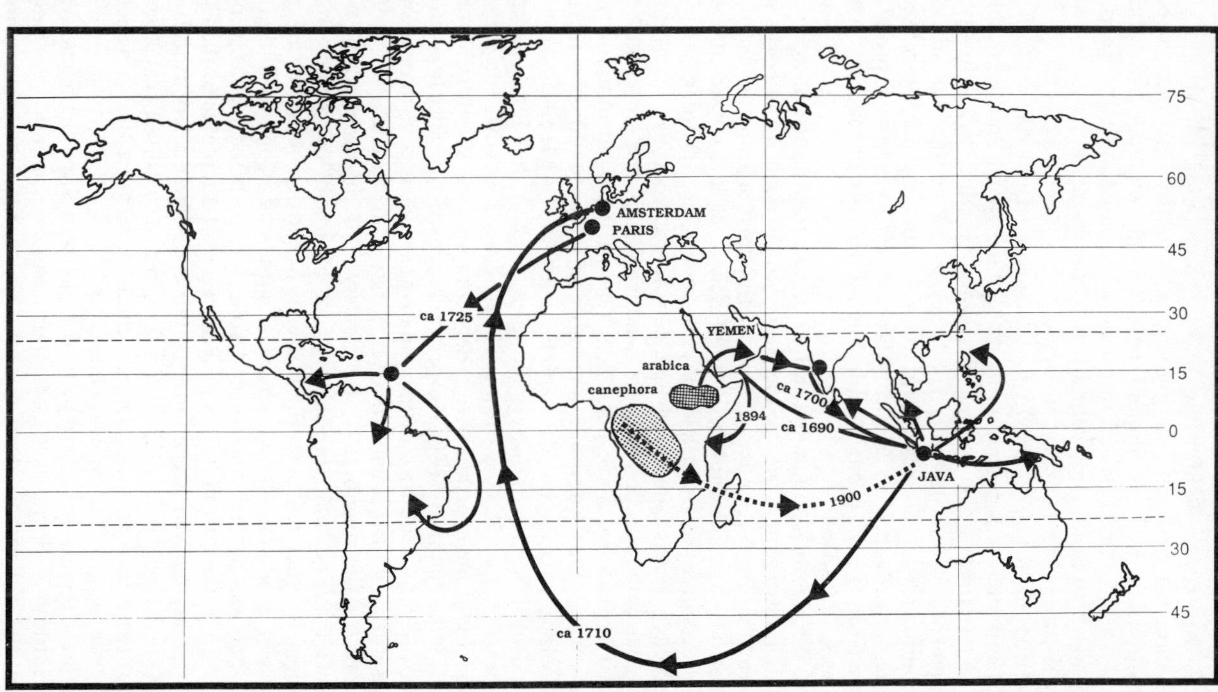

Source : Adapted from Ferwerda, F.P. (1976), "Coffees", in Simmonds, N.W., ed., Evolution of Crop Plants, Longman, Essex, UK.

collection. The publication of Malthus' work on population created panic among the landed gentry; the fear of population outstripping food production led to interest in the use of scientific methods to bolster production as well as import new genetic material for domestic agriculture. The fear of food shortages was compounded by Napoleon's threat to starve Britain into submission by blockading the country, as well as by the loss of American colonies.

The combined fear of defeat by France and competing with the very poor for limited food resources led the landed gentry and the nobility to seek alternative ways to increase food production. Already, the government had pushed up food prices by buying up stocks for distribution to its allies and for its own supplies during subsequent struggles with Napoleon. High food prices led to increased production; marginal lands were brought under the plough and production in arable land was intensified.

These conditions stimulated technological innovations in agricultural production as farmers tried to increase both the quantity and diversity of the available food resources.[21] One of the most significant institutional innovations was the mobilization of scientific knowledge and exotic genetic resources to boost agricultural production. In order to facilitate the process, it was considered necessary to form an institution through which the mobilization would be conducted; this was the origin of the Royal Horticultural Society in 1804. One of the main initiators of the Society was John Wedgwood, a member of the prestigious Staffordshire pottery family. He included Joseph Banks, who was a long-time president of the Royal Society and horticultural adviser to the king, as a charter member. The Earl of Dartmouth was elected as the Society's first president and membership was largely composed of the British nobility.

The Horticultural Society's original prospectus clearly stated its objectives: 'Horticulture . . . may be divided into two distinct branches, the useful and the ornamental; the first must occupy the principle attention of the members of the Society, but the second will not be neglected.'[22] The main objectives were obviously economic. By then British and other travellers had already identified most of the major sources of economic plants; botanists had already done the initial work and what was needed was to send collectors to bring back the plants. Earlier collections from Africa proved to be not adaptable to the English climate and required greenhouse conditions. The Society's decision to send David Douglas in 1823 to North America to collect fruit trees was part of the efforts to introduce new material into the British agriculture and move away from tropical collections.[23]

Before Douglas was dispatched to North America, the agenda of the Horticultural Society was already shifting from economic to ornamental plants. Several factors accounts for this change. First, there was no evidence of the Malthusian nightmare of population overrunning food production. Innovations had partially responded to the challenges of agricultural production. Second, the defeat and banishment of Napoleon removed the fear of a blockade and the landed gentry found it less attractive to engage in scientific gardening, and this was partly because the fear was alleviated by the prospects of sending colonists to the Antipodes. Their interests shifted from increasing food production to adorning their gardens with exotic plants.

By the early 1820s the fellowship of the Horticultural Society had risen to 1,500 and included ardent ornamentalists such as the Emperor of Russia and the kings of Denmark, Bavaria and the Netherlands. With the external pressures seemingly removed and the internal composition of the Society changed, it was almost inevitable that its focus would shift. But despite the change, the organization became a major source of innovation and played a crucial role in supporting plant collection, especially with the technical expertise provided by the Royal Botanic Gardens at Kew.

The plant collections of the 16th and 17th centuries were associated with rapid growth of knowledge on plant propagation and care. New innovations and skills emerged. The construction of propagation houses and covered gardens made it possible to grow plants from almost anywhere in the world. It should be noted that since the early introductions were largely from the temperate region with relatively similar climatic conditions to Britain, the growth of knowledge in plant propagation could easily benefit from previous experience. British gardeners became increasingly skilful in the art, but they could only work with the material that had survived the long journeys from different parts of the world. Some plant hunters had discovered that in the humid tropics it was difficult to collect seed or successfully dry it in order to germinate under glasshouse conditions. In other cases, the arrival of plant hunters did not coincide with the seed maturation period. Moreover, many plants could not be propagated by seed and it was, therefore, necessary to collect and transport plants to Britain.

No effective methods had, however, been devised to transport plants over long distances and through varying temperatures and the records are full of accounts of massive losses during transportation. Thus the imbalance created by the low rate of plant survival and the high growth rate in gardening knowledge (supported by the English nobility) could only be redressed either through speedy transportation or new ways of raising the survival rate of the plants. The first option was not viable as winds still remained the only source of energy for sea transport. Those concerned with the introduction of plants therefore focused on ways of ensuring their survival.

The Horticultural Society that was responsible for many plant collection missions started to work on ways of improving plant transportation in the 1820s. The old plant cabins, rough wooden boxes with tough glass were the technologies of the day. They were accompanied by instructions on how to open, shut, air and water the plants over thousands of miles of land and sea. It was indeed a trial and error process. The Society became concerned by the rate of plant losses and much thought was given to improving the technology. The improvements, however, were only minor refinements of the old techniques.

It should be noted that over this period the English nobility was increasingly interested in plant collection. The Duke of Devonshire, for example, hankered after exotic, rare and flamboyant plants. His garden, like those of other members of the nobility, was adorned with nature's novelty from different parts of the world. This was also a period of rapid urbanization in Britain, and urban folk, removed from the splendour of the countryside, planted exotic plants wherever they could find space. House plants became big business and new firms emerged to promote the industry. The Window Tax in England forced some of the urban dwellers to bring a part of nature's novelty into their houses. The demand for exotic plants increased as the nobility expanded the size of their botanic gardens. Under such conditions, innovations to facilitate the transportation of plants were clearly needed. Improvements on the old methods had reached limits and new ways had to be found.

It was Nathaniel Bagshaw Ward, an amateur naturalist and a general practitioner in the dockland of East London who made the breakthrough. In the summer of 1829 Ward was examining a wide-mouthed bottle covered with a lid in which he had buried the chrysalis of a Sphinx butterfly. After two weeks Ward noticed that fern and grass seedlings had sprouted in the bottle.[24] He argued that the lidded bottle allowed the moisture to rise and condense, thereby maintaining the same degree of humidity. This is the concept behind what is now called the Wardian case or terrarium.[25] The plants survived for four years and died only after the lid had rusted while Ward was away on holiday.

Ward's discovery is treated in the literature as a technological accident.[26] This interpretation tends to ignore the process of social and technological evolution that made the innovation relevant. It was easy for Ward to consider the discovery as a solution to the plant transportation problem. Having been to Jamaica at the age of 13 to collect plants and insects, he was aware of the problems of transporting plants over long distances. His interest in plants went beyond his activities as an amateur naturalist; his profession relied largely on plants as a source of medicine.[27] In addition, Ward flourished in a period of increasing environmental pollution in London due to the rise of industrial manufacturing.[28] His activities included encouraging window-gardening among the working class.[29]

Ward was also in touch with such leading botanists of the day as W. Anderson of the Chelsea Botanic Garden, Sir William J. Hooker of Kew Gardens and John Lindley of the Royal Horticultural Society and was therefore aware of the difficulties of transporting plants over long distances. He had also spoken to the surgeon-naturalist, Archibold Menzies, who had lost all the plants collected with Captain George Vancouver on his voyage around the world. With his knowledge of the problem and his association with the botanic gardens, Ward decided to test out the concept by experimenting with some of the most difficult of all plants under cultivation, the filmy fern (*Trichomanes radicans*). After the initial success with the filmy fern, he tried out other difficult species such as *Aspidium molle*, date palm (*Phoenix dactylifera*), *Rhapis flabelliformis*, *Dendrobium pulchellum*, and *Mammillaria tenuis*.[30] In 1832, he filled two cases with ferns, mosses, roses and violets to send to Australia, where the immigrants bemoaned the absence of the familiar English greenery.[31] The cases were modelled on the concept of the bottle but shaped like the old plant cabins and were hermetically sealed, making them miniature airtight greenhouses. They survived a gruelling eight months through varying temperatures, heavy seas, storms, sunshine and decks awash with salt. In appreciation, Ward's Australian friends filled the cases with local plants and sent them to him.[32]

Inspired by the prospects of the Wardian case, the Sixth Duke of Devonshire and his gardener, Joseph Paxton, set out to enrich their gardens with exotic plants of Asia. In 1835, they dispatched 24-year-old John Gibson, whose botanic treasure hunt centred on the search for the fabulous tree of Ind (*Amherstia nobilis*); this was the largest botanic expedition hitherto mounted. The expedition started with a collection of English plants to be presented to Nathaniel Wallich (Nathan Wolff) at his Calcutta botanic garden. Gibson left India with two amherstias, one for the Duke and another from Wallich to the Court of Directors of the Honourable East India Company, his employers. The plant addressed to the Duke was almost the only one that died during the homebound trip and the directors allowed him to pass their plant on to him.[33]

The impact of the innovation was immense. Before the introduction of the cases, plant collectors lost an average of 99.9 per cent of all plants shipped from China to England. In the 1840s, the Wardian cases had reduced the average loss to 14 per cent. Conrad Loddiges and Sons, then London specialists in exotic plants, reduced their losses from 19 to one out of 20 cases with the use of Wardian cases. William J. Hooker imported to Kew Botanic Gardens six times as many exotic plants in 15 years as had been sent in the entire previous century. Not only did the technology become a medium for plant exchange across the Atlantic, but it stimulated the growth of the horticultural industry, as new and rare plants could now be transported from various parts of the world.[34]

In addition to adorning the Victorian living rooms, this seemingly modest innovation was later to revolutionize plant transportation across the world and help strengthen the British empire by the introduction of new agricultural crops in its colonies. Among other parts of the world Ward was commissioned to send plants in his cases to Alexandria, Damascus, Calcutta and Pará. The cases were immediately adopted for transporting economic plants to different parts of the world. One of the earliest major applications was undertaken by John Williams, a missionary who was later known as 'the martyr of Erromanga'. He used the cases to introduce Chinese bananas in Samoa, from where George Pritchard took them to Tonga and Fiji in 1840.[35] The Wardian case enabled the British to move tea (*Camellia sinensis*) from China to India, rubber (*Hevea* species) from Brazil to Ceylon and *Cinchona* from Latin America to India. On the whole, the British empire expanded largely as a result of the application of botanical knowledge, technical change and institutional organization to agricultural production.

Genetic resources and the British Empire

The history of the British Empire is closely associated with the movement of plants from one country to another although conventional historical texts have tended to emphasize the role of land acquisition and exploitation of labour in the colonies. The role of genetic resources in colonial expansion is often ignored. The economic pressures that led to the discovery of the Americas were part of a dynamic and complex process of economic change in Europe. The search for knowledge and new plants was already part of the culture during the early period of colonial expansion and imperialism. The role of genetic resources in the rise of the British Empire is an example of this process and the imperatives that led to the redrawing of the global genetic map.

The practice of plant transfer associated with Columbus' travels continued into the colonial period. Plants that had been domesticated in the nearby regions were transferred farther afield to serve the interests of colonial agriculture. Let us examine the case of tea transfer from China to India, which was the first major attempt by the British to establish an economic base in a major exotic plant in the colonies.[36] The process was a result of intimate links between international trade, plant exploration and economic complexity. Plant exploration in China started in the late 17th century with the pioneering work of James Cuninghame, a surgeon with the East India Company.[37] The floral similarities between China and Europe made plant exploration in the region a favourable endeavour. Certain genera such as Magnolia, Hamamelis and Wisteria occur only in eastern USA, Japan and China. Apart from North America, China has made the most remarkable contribution to British gardens.[38] The early collections were later to become the beginning of significant developments in historical botany.

Britain was a major importer of Chinese tea,[39] earning the British government an estimated £3.0 million in customs duties annually, nearly 50 per cent of the budget of the Royal Navy.[40] An early attempt to increase foreign earnings for Britain was to force China into the opium trade and, when the Chinese authorities attempted to stop this trade, British troops moved in and the Opium War (1839-42) was started. The war not only helped open up large sections of China to trade and plant exploration, but it also weakened its monopoly in the tea trade. One way of reducing the monopoly was to move tea production to a British colony.

In 1827 and 1834, a British East India Company botanist and superintendent of the Saharanpore Botanic Garden reported that the foothills of the Himalayas would be suitable for tea production.[41] Initial attempts to transplant tea from the Calcutta Botanic Garden to the region failed. In 1848, some 11 months after the signing of the Treaty of Nankin, the company commissioned Robert Fortune to undertake the task of relocating tea. His mission included the collection of information on tea cultivation. He brought about 2,000 tea plants and 17,000 seeds to India as well as Chinese experts in tea cultivation, and was thus able to establish tea plantations in Darjeeling and Assam. By the late 19th century, India had already displaced China as the main tea exporter to Britain. Later tea was also planted in Ceylon (Sri Lanka) and moved to several other British colonies with suitable conditions such as Java and to the present Kenya, Tanzania and Malawi. It is interesting to note that the indigenous Assamese tea variety (*Camellia sinensis* var. *assamica)* was cleared to allow for the plantation of Chinese trees. The imported trees either died or were unproductive and later the hills had to be searched for wild varieties to cross-breed with the imported ones.

The process of tea seed collection was a joint effort between the government, scientific community and the private sector. The delay in the transfer of tea to India, and continued control over the production knowledge by China, were partly a result of the false belief that 'green' and 'black' tea were from different types of plants; a confusion exacerbated by Linnaeus' misclassification.[42] Chinese control over the knowledge required to establish tea plantations was, however, weakened after their military defeat. The transfer of tea to India was facilitated with the systematic involvement of Kew Gardens. At this time, botanic gardens were important agents for the introduction of economic plants in the colonies. The transfer of rubber *(Hevea brasiliensis)* from Brazil to South-East Asia was also undertaken by Kew Gardens; other crops established through botanic gardens include the African oil palm *(Elaeis guineensis)* and breadfruit. But by the late 1800s, plant introduction and propagation had been taken over by agricultural research institutions.[43]

Cinchona *(Cinchona* species) was one of the major plants transferred from the Andes to India and other parts of the world by the British Empire.[44] The plant yields quinine which is used as a cure for malaria. The alkaloids in the plant interfere with the reproduction of the malaria parasite. Cinchona not only had humanitarian benefits but military, political and economic implications. For the Empire to be expanded and consolidated, the effects of malaria had to be reduced, and cinchona was the main remedy. Armed with the experience of the tea transfer and Wardian cases, John Forbes Royle of the East India Medical Board shipped six young plants of *C. calisaya* from Kew and Edinburgh Botanic Gardens to India; they died at Calcutta. It was later agreed between colonial officers and botanists at the Calcutta Botanic Garden that the only solution was to send an expedition to the Andes. Funded by the India Office and organized with the help of Kew, collectors were dispatched to the Andes. On 31 December 1860 Richard Spruce and Robert Cross shipped 100,000 cinchona seeds and 637 young plants (mainly of *C. succirubra)* to England and subsequently to India.[45]

The collection, especially from Bolivia, contravened national laws, as cinchona export was a government monopoly and after Cross and Spruce had left, a law banning the export of plants was introduced in Ecuador in May 1861. Cross however, returned to the country later that year and collected more seed, despite being aware of the new legislation; other members of the team also continued to collect cinchona from the Andes. Over the period 1868-70, Cross collected more cinchona in Colombia, including *C. pitayensis,* voted as the best variety sent to India.[46] The Nilgiri Hills in India became the home of the new plants and cultivation began. Before large-scale production started, major breeding and seed exchange projects were initiated. It was through the seed exchange period that the Dutch obtained some of the seeds they used to establish cinchona plantations in Java. This became a major source of quinine, not only due to the use of a relatively superior cinchona species but also because of high rainfall. It is important to note that the British were primarily interested in protecting their troops while the Dutch grew cinchona for commercial purposes.[47]

Not only did the transfer of cinchona reduce the market share of Latin American exporters but it also contributed to the expansion of colonialism in Africa. The consolidation of the British companies in West Africa and subsequent expansion of colonial commercial interests would have been more difficult without the help of quinine. Malaria had made the navigation of the Niger River extremely difficult and dangerous. It is interesting that quinine was available either as totaquine or quinine sulphate, although the latter was chosen and widely used because it yielded higher profit.[48] On the whole, the search for plants, the rise of applied botany and advances in evolutionary theory were part of complex interrelationships between government departments, scientific institutions, trading companies and military operations. These interrelationships were associated with the growing complexity of the British economy and their internationalization. The acquisition of colonies was not enough unless linked with the availability of labour and plant genetic resources. The relocation of plants was therefore linked to complex institutional relationships which evolved to fulfil the requirements of colonial agriculture.[49] The details of the British use of genetic resources for imperialist expansion was specific to that country and cannot be applied to explain what happened in the US. What is important, however, is the theme that economic transformation over the period relied heavily on the introduction of new genetic material and the related technologies into the economic system.

US historical botany

The role of plant collection in the rise of the British Empire illustrates the combined efforts of the government, scientific institutions and dedicated individuals to shape the direction of social evolution through the introduction into the economy of new genetic material and knowledge. These pioneering efforts were undertaken largely in the context of colonial imperatives. But as argued earlier, history is non-linear and some evolutionary processes manifest themselves in different forms depending on the prevailing internal and external circumstances. While genetic resources helped Britain expand the Empire, similar material was used to build a domestic agricultural economy in the US that stands unrivalled. The agricultural model that evolved from the process was later to shape the current agricultural policies of most of the Third World.

The US is undisputedly the world's agricultural superpower. It is largely the strength of the US agriculture that has given the country the vast capacity to industrialize and the economic strength to influence the rest of the world. North America is relatively poor in indigenous crop genetic resources,[50] its main indigenous contributions to global economic botany include blueberry (*Vaccinium atrococcus*), cranberry (*Vaccinium macrocarpon*), Jerusalem artichoke (*Helianthus tuberosum*), pecan (*Carya illinoensis*), sunflower (*Helianthus annuus*), black walnut (*Juglans nigra*), black raspberry (*Rubus occidentalis*), American red raspberry (*Rubus strigosus*), muscadine grape (*Vitis rotundifolia*), American wild gooseberry, wheatgrass and tufted hairgrass (*Deschampsia cespitosa*).[51] How then did a continent of berries become a global agricultural power? The answer lies in the capacity to introduce into the country new plants and the related agricultural know-how and allow for relative autonomy and diversity in experimentation. The history of US agriculture is largely a study of plant introduction, technical change and institutional reform.

The first major influx of genetic material into the US was associated with what Crosby calls the Columbian exchange.[52] By the time of the Columbian explorations, Europe had accumulated a large stock of genetic resources from other parts of the Old World. On his first voyage, Columbus left seeds with the citizens of the abortive Navidad colony in 1493; the fate of the seeds is not known because the citizens were killed by the Arawaks. On his second voyage, Columbus brought to Española livestock and seeds of wheat (*Triticum* species), chickpeas (*Cicer arietinum*), melons (*Cucumis* species), onions (*Allium* species), radishes (*Raphanus sativa*), salad greens, grape vines (*Vitis* species), fruit stones for orchards and sugar-cane (*Saccharum officinarum*) from the Canaries.[53] The early phase of colonial settlement was marked by intensive plant trials so as to re-establish the European dishes and way of life. It was not only experimentation that characterized the period – the Spaniards searched South America for areas where crops such as wheat and vines could grow. The frontiers were pushed well into the Andes in search of possible sources of the main ingredients of the Castilian lifestyle – bread, wine and olive oil.[54]

Under the classical colonialism of the Spaniards, the South American Indians were forced to grow wheat and other European crops either for the colonists or to make tribute payment in kind. The Indians hardly added these exotic plants to their menu; similarly, most colonists did not abandon their Castilian tastes and adapt to local staples. In order to maintain the supremacy of exotic plants and guarantee the availability of native labour to produce them, the Spandiards launched vicious campaigns to undermine the cultivation of such indigenous crops as amaranth. The plant was so tightly integrated into the local culture that its forceful removal from public life contributed to the destruction of local religions and cultural practices.

By 1550, the ruthless exploitation of local labour in commercial plantations, and the newly-introduced diseases such as smallpox, had destroyed the indigenous Española population. Their fellow Arawaks in Cuba, Puerto Rico and Jamaica were later quickly consigned to oblivion. Where the Spanish did not settle, slavers spread diseases and raided the islands to collect slaves for Española, Cuba, Puerto Rico and Jamaica. As the islands were depopulated, the number of horses, dogs, pigs, cattle, chickens, sheep, and goats increased rapidly; these animals had no predators and were affected by few or no local diseases. 'Their number burgeoned so rapidly, in fact, that doubtlessly they had much to do with the extinction of certain plants, animals, and even the Indians themselves, whose gardens they encroached upon.'[55] It was the build-up of animal populations on the islands, in addition to smallpox, that provided the biological arsenal with which the conquistadors stormed the mainland.[56]

The Spaniards introduced figs, dates, grapes, olives (*Olea europaea*) and pomegranates (*Punica granatum*) in their missions in New Mexico and California. They also introduced alfalfa (*Mendicago sativa*), lemons (*Citrus limon*), oranges, and ginger (*Zingiber* species). Many of these had originated in Asia, reached Europe and finally found their way across the Atlantic. The post-Columbian era turned North America into a large testing ground for new plants. Although the Europeans raised American plants such as tobacco (*Nicotiana tabacum*), cocoa (*Theobroma cacao*), paprika (*Capsicum annuum*), and cotton (*Gossypium barbadense*) on large plantations, the main transformation of the landscape resulted from the introduction of Old World plants and animals.

This transformation is 'probably the greatest biological revolution in the Americas since the end of the Pleistocene era'.[57] This irreversible process was accomplished by strewing seeds all across the continent. Wherever the pathfinders set foot, they planted Old World seeds and observed their performance. Jacques Cartier, who brought back to France two new conifers from Canada, recorded that on his 1541 voyage, his men planted European cabbage (*Brassica oleracea*), lettuce (*Lactuca* species), and turnips (*Brassica rapa*). On his 1583 expedition to Newfoundland, Sir Humphrey Gilbert sowed and harvested peas, and in the late 16th century English fishermen planted wheat, barley, rye, oats, beans, peas, kernels, plums, nuts and herbs in Newfoundland where they reportedly prospered as in England.

Colonial expeditions to America were incomplete without a catalogue of plants to be tried out. The Endicott expedition of 1628 for the Massachusetts Bay Colony, for example, was directed to take with it seeds of wheat, rye (*Secale cereale*), barley (*Hordeum vulgare*), oats (*Avena* species), beans, peas, stones of peaches (*Prunus persica*), plums (*Prunus domestica*), cherries, and seeds of filberts (*Corylus* species), pears (*Pyrus* species), apples (*Malus domestica*), quince (*Cydonia oblonga*), and pomegranate, woad seed (*Isatis tinctoria*), saffron (*Crocus sativus*), licorice (*Glycyrrhiza* species), madder (*Rubia tinctorum*), potatoes (*Solanum tuberosum*), hop (*Humulus lupulus*), hemp (*Canabis sativa*), flax (*Linum usitatissimum*), and currant plants (*Ribes* species). This array of plants was aimed at starting an agricultural economy based on crops known in England. By 1630 the gardens of Massachusetts were already growing beans, peas, onions, radishes, spinach (*Spinacia oleracea*), lettuce, turnips and cabbage.

During the early colonial period other cereals, such as rice (*Oryza sativa*), and pasture crops were introduced into the US. Crops such as pearl millet (*Pennisetum glaucum*) and sorghum (*Sorghum* species) which were brought into the US from Africa by slaves started to spread over this period. Pearl millet was first noticed growing on Negro plantations in

Jamaica and was later grown by slaves in the South. The common millet (*Pennisetum miliaceum*) grown in Massachussets in 1637 became a major source of chicken feed a century later. By the 1660s other plants of African origin, such as cowpeas, were growing in New England, and in South Carolina by the early 1680s. By the end of the 17th century, numerous crops from all over the world had reached the US. Crops of West Indian origin such as cotton later entered the US economic system.

The late 17th century saw the early stages of institutional reorganization to respond to the increasing demand for imported plants and the generation of new knowledge on plant propagation. One of the earliest public efforts to establish an experimental farm was in 1699 on the banks of the Ashley River in South Carolina. This farm, set up by Lords Proprietors, tested the adaptability of various crops, and represented an effort to organize the generation of knowledge and identify critical features which influenced the adaptability of imported plants. The founders of the farm recommended that vine, hemp, indigo (*Indigofera* species), tobacco, silk, flax and ginger be produced for export.

Other early public sector efforts included the founding of the Trustee's Garden of Georgia at Savannah in 1733. The experimental farm was designed in England before the colonization of the region and was planned to help in setting up a silk production centre. The area was chosen because of the availability of indigenous mulberry trees. Over the same period, individual efforts were also leading to major innovations in economic botany. In 1730, John Bartram started the first botanic garden in the US, on the banks of the Schuylkill River three miles above Philadelphia. An avid plant collector, Bartram travelled widely to study American flora, selling seeds and plants to support his work. At Flushing, Long Island, later in the same year, the Linnean Botanic Garden was founded which tried to acquire foreign and local plants, especially grapes. Later it became a commercial farm under the Prince family. Its efforts included the introduction and popularization of new plants.

Britain saw the American colonies as a potential source of economic gains and therefore encouraged the transfer of plants collected in other countries to the US. In 1770, John Ellis published a book with instructions on how to transport plants over long distances. Since China was considered a major source of economic plants, its distance from Britain and America made it necessary to review the existing methods of transporting plants. Missionaries were considered the main sources of information on such plants and how to acquire them. The best source for plants, however, were the resident-factors in China, such as John Bradby Blake, who brought upland rice to South Carolina and other useful plants such as the Chinese rhubarb (*Rheum officinale*) from Canton.

The US's botanic history has two complementary strands which later contributed to the country's rise as a global agricultural power. First, it was one of the first regions to benefit from the botanic knowledge that English gardeners and plant hunters had accumulated over the previous century. Second, and even more important, was the capacity of the colonists to adapt the knowledge to the local conditions and to initiate independent experiments. This drift towards independence was also associated with broader movements by the colonists to assert their freedom from Britain. Thus some of the botanic gardens established to serve colonial interests ceased to function during the Revolutionary War. The US was gradually reducing its dependence on external sources of agricultural knowledge and becoming a major innovator in its own right.

US citizens initiated global searches for plants of economic value and the related knowledge. The search was intensified in the 1760s as local agricultural production started to expand. Prominent US representatives abroad became agents of plant introduction. Benjamin Franklin, for example, sent home seeds of various plants while in England as an agent of the colony of Pennsylvania over the period 1764-75. Some of his seed collections and agricultural information were sent to John Bartram, the founder of the first US botanic garden. Franklin is associated with the introduction into the US of rhubarb (*Rheum rhabarbarum*), upland rice, broom corn and soybean (*Glycine max*).

Franklin was not the only US statesman to promote plant introduction. The early history of the US is characterized by people who can be referred to as 'botanist-kings'; the likes of Franklin, Washington and Jefferson. George Washington not only requested English agriculturalists to procure seeds for him, but he also made the first recommendation that a section of the National Government be set up to cater for the interests of farmers. Like Franklin, Washington recognized the need for dependence on England for the supply of seed while at the same time undertaking local experimentation. But he went further and suggested institutional support for the process of plant introduction.

The tradition of 'botanist-king' was carried further by Thomas Jefferson who believed that it was extremely important for young nations such as the US to undertake agricultural experiments. Jefferson is remembered for his widely-quoted statement: 'The greatest service which can be rendered to any country is to add a useful plant to its culture.'[58] He conceived the US as an adaptive terrain that had not fully realized its potential since it did not possess the articles of culture for which nature had endowed the country. He argued that the process of discovering these articles would require extensive experimentation which would entail abundant failures; any successful results would repay the effort. Unlike Washington, he saw this not as the role of the government but of agricultural and botanic associations.[59] He also collected seed from all over the world and was actively involved in the introduction of rice and olive trees into the US.

The era of 'botanist-kings' was closely associated with the rise of agricultural societies promoting the introduction of new plants. These societies were involved mainly in disseminating agricultural information and exchanging seeds. *Agricultural Museum,* the first agricultural periodical in the US, reported mainly on the activities of the societies. The South Carolina Agricultural Society pioneered the introduction of exotic plants of economic value and, partly as a result of surplus cotton in the state, the Society appointed a committee to look into the benefits of imported plants. In 1823, this committee noted that the state had benefited from imported crops such as rice, indigo and cotton and recommended that further introductions be encouraged. The Society later appointed a committee of three to work closely with consuls of the US, other persons in foreign countries and the navy, to introduce staples that might substitute for cotton. Some US$200 was set aside annually to support the efforts; the resulting seeds and information were distributed free to members.[60]

The societies were largely adaptive networks of individuals who came together to realize particular objectives. At this time no effective federal institutions existed to perform these functions; individual efforts, often in social networks, were the main agents of plant introduction. The last major pre-Federal efforts to import plants were undertaken by Henry Perrine, a physician, who dealt mainly with tropical plants. Perrine's work was aimed mainly at diversifying and strengthening the economy of the South, especially in the newly-acquired Territory of Florida. He was appointed a US consul at Campeche, Mexico in 1827 and started immediately exporting Mexican plants, particularly *Agave* species for producing fibre.[61] In 1838 the US Congress granted Perrine land (23,040 acres) at Indian Key, Florida; this was the last land grant made by Congress to encourage plant introduction.[62] It became increasingly obvious that individual efforts and farmers' associations could not cope with the demands and complexity of the growing US agriculture and the related uncertainties. Federal assistance became necessary.

Congress had no experience in supporting large plant introduction projects. In 1802, the first attempt involved the authorization of John Dufour and his associates to purchase land north-west of the Ohio River between the Great Miami River and the Indian boundary line. The project was aimed at producing wine for the US market, but after 16 years of experimentation and Congress extensions on the land, the project failed. A similar Congress grant to French emigrants, known as the Tombigbee Association, failed to yield wine and olive oil. The association settled in western Alabama in 1819 and was granted 92,160 acres. The colony was unable to supply its food requirements, fend off violent squatters, provide enough labour for clearing the land, and receive cuttings in season.

Having reached limits to the use of the genetic material easily available in Britain and other European countries, the Federal Government sought to intensify the search process through the mobilization of its various state organs. The Secretary of the Treasury, William Crawford, issued a circular in 1819 which was aimed at encouraging naval officers and consular officials to send home any plants deemed useful to the US agriculture.[63] The appeal did not yield much since no finances were allocated to the exercise. The only result was the importation of barrels of *lupinella* from Consul Appleton in Italy.

In order to collect plants without any funding, the Federal government had to identify institutions that were in a position to undertake such missions while carrying out their normal duties. The navy, which was not involved at the time in any major military expeditions, was identified as the most suitable institution for undertaking the task. In 1827 another Treasury circular was issued which contained more details on methods of propagation, cultivation, uses, preferred soil and climate as well as transportation of the seeds and plants; this was directed more specifically at the navy. The circular, masterminded by the Secretary of the Treasury, John Quincy Adams, did not authorize any funds but was issued in close collaboration with the navy and the Department of State. The seeds and plants were shipped to a government botanic garden in Washington from where they were distributed.[64]

Since there was no special Federal office in charge of plant collection, the navy became the most active department in rendering this service, and kept a squadron in the Mediterranean from where numerous plants were shipped to the US.[65] In the absence of clear guidelines for accountability, captains of merchant and naval vessels collected plants and livestock for their own farms. Some officers even sacrificed the comfort of their crew to give room to livestock. One Captain Jesse Elliot was court martialled for the offence and a general order was issued prohibiting the transportation of animals on public ships.[66]

A House of Representatives resolution passed in 1830 requested the navy and officials in foreign countries to help in securing new sugar-cane varieties and other economic plants suitable for the US climate and soil. In response, the West Indian Squadron collected several varieties in Trinidad and brought them to Pensacola; they were distributed by the governor of Florida. With backing from the House, it was possible for the navy to launch full-fledged botanic explorations using their expertise as well as professional botanists. At this moment in the history of the US, the interpretation of national security clearly encompassed the need to respond to the growing complexity of agriculture and its links with the rest of the economy. The navy's involvement was, therefore, relevant to the overall objectives of the national economic policy.

It was in this spirit that the navy launched a botanic expedition to the

Pacific under Captain Charles Wilkes in 1838, with the botanist, William Rich as part of the team. Charles Wilkes and his men were on this journey for four years, dispatching numerous seed packets and Wardian cases from Madeira, St Iago, Rio de Janeiro, Fiji, Australia and Hawaii. The seeds sent to the US were distributed to farmers, other material was deposited in a greenhouse in Washington, later called the Botanic Garden. In subsequent years, the navy was mobilized under emergency-like conditions to collect seed for boosting existing agricultural production. For example, in 1852, when a Louisiana farmer noticed that sugar-cane in the state was degenerating, the Secretary of the Navy ordered the East Indian Squadron to secure suger-cane varieties, especially the *Salangore* cane (reported to be available at Penang in Malaysia) as well as *Otaheite* and *Mauritius* canes. The Sloop of War *Marion* was held ready to rush the collections back to the US while another ship, *St. Mary's,* was instructed to secure additional *Salangore* cuttings. The first collections decayed but a subsequent attempt was successful.

The fact that international trade, seed collection and naval expeditions were closely linked to the US economic policy is further exemplified by the Perry Naval Expedition of 1853 to Japan, aimed at opening Japan to trade with the US. One of the expedition's activities was to exchange seeds and agricultural technology with the Japanese. On board was James Morrow who was instructed by the Department of State to 'endeavour to introduce those vegetable productions not indigenous at such place of rendezvous . . . and carefully note and collect all indigenous vegetable products . . . with a view to their introduction into the United States, preserving seeds and dried specimens of as many plants as possible.'[67] Morrow was not well prepared for the mission; first he had no money to buy seeds, and when he got the funds, he realized he did not have Wardian cases. Most of his plant collection from Japan perished on its way to China. Despite the problems, the expedition yielded numerous collections of wheat, barley, turnips, rice, persimmons (*Diospyros kaki*), tangerines (*Citrus reticulata*), kumquats (*Fortunella* species), roses, tobacco, sugar-cane, cotton seeds and ornamentals from Japan, China, Java, Malaysia, Mauritius and South Africa.[68]

In all the previous operations, the navy had to attend to their normal duties and plant collection was a subsidiary, although important, activity. The first naval vessel to be sent out on a purely agricultural mission was the *Release* in 1856. With an appropriation of US$10,000 the *Release* was sent to South America by the Patent Office to acquire more cuttings of cane. In 1857, the vessel brought back 1,000 boxes of cane cuttings, plantain, banana, eddo (*Colocasia esculenta*), yam roots (*Dioscorea* species) and other plants. The expedition was initially deemed a success until it was discovered they had also imported into the country the cane borer.[69] The head of the agricultural activities on the *Release,* Townsend Glover, devoted much of his time to curbing the spread of the worm; he later became an entomologist with the Department of Agriculture and taught at Maryland Agricultural College.

Additional plants were added to the US collection by diplomatic missions overseas. John Davis, the US consul at Canton, China, sent seeds to the US in 1849, these had reached Davis from S. Wells Williams, the missionary and linguist who later served as an interpreter for the Perry expedition. The contribution to the US included persimmon, muskmelon (*Cucumis melo*), olive and watermelon seeds (*Citrullus lanatus*). In 1851, the US consulate at Panama sent to Utica, New York, potatoes which were later used to breed new varieties resistant to potato rot. The potato rot epidemic had destroyed large sections of crop during the 1843-47 period.

The Panama consignment helped introduce new genetic material into the cultivated stock and therefore revitalize the US potato crop.[70] US consulates were expected to collect agricultural information in the host countries. Peru and Chile were considered a potential source of wheat, alfalfa, pepper, corn and beans and in order to ensure access to the material, the US encouraged seed exchange with those countries. Seeds were also received from US consuls in India, Turkey, China and Japan. In common with the naval officers, diplomatic officials also sent seed to their friends or brought them home as personal collections hoping to benefit from their introduction.[71]

In the 1810s and 1820s, the US undertook significant institutional reforms to facilitate the introduction of plants. In 1816 Congress revised tariff regulations to allow exotic plants and trees to enter the country duty-free. In 1822, the 200-acre space between the Capitol and the Washington monument was turned into an experimental farm for propagating imported seeds and plants. Not until 1839 did the Federal government appropriate US$1,000 from the Patent Office to support seed collection and distribution activities as well as to publish the related statistics. The activities were conducted under the new Agricultural Department of the Patent Office, which came under the jurisdiction of the State Department. It is notable that long before the Department of Agriculture was formed, the US considered plant introduction part of new discoveries and inventions. This shows that the role of knowledge, technology and genetic material in the economy was well recognized.[72]

The 1840s were characterized by free international exchange of seeds. In 1847 the US distributed more than 60,000 seed packages, some of which were contributed by the French Minister of Agriculture and Commerce through Alexandre Vattemare, advocate of international seed exchange. Foreign consuls based in the US imported seed as part of the exchange programme. The Bavarian consul at Philadelphia, C. F.

Hagedorn, for example, imported seed from his country for distribution in the US. Seed exchange arrangements were made with the governments of Austria, Bavaria, Prussia, China, Japan, India, Guatemala, British Honduras, Brazil, Russia, and Switzerland.[73] As more seeds were distributed, so did reports of poor germination begin to emerge. At this time seed distribution was carried out mainly under the postal franking system.

As the Agricultural Division expanded its activities, it became necessary to move it from the Patent Office to the Department of the Interior, a task accomplished in 1839. In the following years the Agricultural Division sent its own people to foreign countries to buy seeds. By this time, local suppliers of seeds were beginning to form themselves into a nascent industry. They strongly criticized the decision of the Agricultural Division to send D. J. Browne to Europe to buy seeds. Although the seed dealers succeeded in having a Senate investigation committee look into Browne's seed buying activities in Europe, they could not underrate his contributions to the US agriculture.[74] Additional seeds were also sourced from China with, in 1856 alone, some US$1,000 allocated to the procurement of seeds from China.

With its own working budget, the Agricultural Division was also able to make seed supply arrangements with European firms such as Vilmorin-Andrieux (Paris), Charlwood and Cummins (London), Ernest von Spreckelesen (Hamburg), and William Skirving (Liverpool). By 1856 the Congressional allocation to agriculture was US$75,000, a large share of which was spent on seed importation, mostly for common garden plants. This also made it easier for the Agricultural Division to use the franking system because small quantities sufficed for experimentation. Over this period, the government seemed to make it difficult for private seed companies to operate. Some people used government seed for planting their regular vegetable gardens and not for experimental purposes. It was felt that the government should confine itself to the distribution of seeds that had hitherto not been introduced. Much of the current debate on the private control of seeds is rooted in the debates of those days. The curtailment of seed distribution in 1859 led to increased emphasis on specific crops.

Critics of the Agricultural Division argued that the body was no more than an appendage of the Patent Office established to provide Congressmen with cuttings and garden seeds for their favoured constituents. At the same time agricultural societies, farmers, and journals, continued to advocate the formation of a separate agricultural institution that was at par with other federal departments. It took about ten years of campaigning for Congress to set up the Department of Agriculture in 1862. Its objectives were to educate the public through the collection and dissemination of agricultural information, and the collection and distribution of seeds.[75] The Department was headed by a Commissioner and it was not until 1889 that it became a full Department headed by a Secretary enjoying ministerial powers. The establishment of the Department of Agriculture was associated with other major political and institutional changes in the US and followed the secession of the southern states, and the promulgation of the Homestead as well as the Morrill Land Grant College Acts.

The period of the Commissionership (1862-89) was marked by major changes in seed collection and distribution activities. The nascent seed industry continued to put pressure on the government to reduce its competition with seed dealers. Instead of collecting and distributing new crops, the first commissioner, Isaac Newton of Pennsylvania, was more interested in better varieties of already established crops, and this is what brought him in conflict with the seed industry. He made his job more difficult by emphasizing the agricultural independence of the South. To achieve this, he stressed the importation of tropical and sub-tropical crops. While the emerging tendency of the established farmers was to move into large-scale production, Newton underscored the role of small farmers and a motivated labour force. His death from sunstroke on a farm in 1867 pre-empted his dismissal by President Johnson. Seed distribution was curtailed after his death.

The case for crop diversification became more obvious in the late 1860s. In the first place, strengthening agricultural production in the South required the introduction of new crops as well as the development of new varieties of existing ones. Second, the US agriculture was already experiencing overproduction which reduced the prices of some staples. The persistent use of the same crop was starting to reduce soil fertility and other problems of monocultural production were becoming more obvious. The potato rot epidemic had illustrated the risks of depending on a narrow genetic base; reductions in sugar-cane yields had necessitated repeated introductions of new varieties. Genetic diversity could allow crop rotation and therefore make full use of the available labour.

Significant shifts in seed collection and distribution emerged in the 1870s, starting with the Commissionership of Frederick Watts (1871-77) through that of William Le Duc (1877-81), George Loring (1881-85) and Norman Colman (1885-89). Over this period, the Department of Agriculture started to buy more seed from local dealers and to spend more resources on importing tropical and sub-tropical crops. Under Le Duc, the extensive distribution of seeds was curtailed and agricultural societies were considered more suitable avenues for plant trial; the number of seeds distributed through Congressmen was reduced and the system finally stopped in 1923. Under Colman, the distribution of seeds declined while the search for new crops was intensified. It also became necessary to introduce quality control in seed distribution, therefore new tests for quality, germination, and disease infestation were introduced as well as requirements for proper labelling. Tests were undertaken to ensure that old varieties were not sold as new.

Table 2.1
Federal germplasm distribution, 1862–1897

Year	Seed Packages	Plants and Cuttings	Annual Budget ($)
1862	306,304	n.a.	—
1863	1,200,000	25,750	—
1864	1,000,000	30,000	—
1865	763,231	35,000	—
1866	992,062	34,000	—
1867	1,426,637	42,125	—
1868	592,398	30,000	—
1869	317,347	31,700	—
1870	358,391	n.a.	25,000
1871	647,321	n.a.	45,000
1872	814,565	n.a.	45,000
1873	1,050,886	n.a.	55,000
1874	1,286,335	n.a.	65,000
1875	2,221,532	n.a.	65,000
1876	1,520,207	n.a.	65,000
1877	2,333,474	n.a.	65,000
1878	1,115,886	57,155	75,000
1879	1,545,739	36,673	75,000
1880	1,581,253	156,862	75,000
1881	1,878,772	100,000	80,000
1882	2,396,476	70,000	80,000
1883	2,467,230	n.a.	80,000
1884	3,622,738	100,000	75,000
1885	4,667,826	74,000	100,000
1886	4,264,165	n.a.	100,000
1887	4,561,741	n.a.	100,000
1888	4,655,519	n.a.	100,000
1889	4,852,512	45,000	104,200
1890	5,605,246	80,000	104,200
1891	6,013,613	117,000	105,400
1892	5,932,989	66,000	105,400
1893	7,704,943	60,000	129,637
1894	9,555,318	75,000	135,400
1895	9,528,632	73,485	148,830
1896	12,000,000	n.a.	185,400
1897	20,368,724	56,100	150,000

n.a. = no data available.
Source: Klose, *America's Crop Heritage.*

The US agricultural advancement was largely an effort to adapt to new as well as changing conditions by applying new technological knowledge or introducing new genetic material. This was also the case for forestry. As settlers moved westward into the plains, deforestation and soil degradation became a major problem,[76] in response a Congressional act of 1880 set aside land for conservation and reforestation. But knowledge of viable tree species was meagre and little progress was made. Moreover, the long-term requirement of reforestation and soil conservation put excessive demands on the population. In 1897, seeds of the Australian tan bark wattle (*Acacia mearnsi*) were mailed to residents on the Gulf, Pacific Coast and south-west states. A prominent botanist, Edward Palmer, was sent to Mexico the same year to collect dryland tree species for the south-west.

The end of the 19th century saw increased tension between the need to consolidate established crops and to introduce new ones. Advocates of established crops wanted to optimize their gains from known genetic material, especially after the Federal government had underwritten the initial costs of plant collection and experimentation. At this stage, the public sector was seen as an obstacle to agricultural advancement because it competed with the nascent seed industry. Another group of people who still wanted to search the world for new plants and try them out advocated increased government spending on seed imports. By then, advocates of established plants were already gaining ground and the tension has never been resolved, indeed, since then US historical botany has been partly a persistent attempt to reconcile the two positions.[77] At the time, the virtual complementarity between the two positions was not well recognized because of limited knowledge of genetics and the process of agricultural evolution.

By the turn of the century, crops such as wheat, oat, rye, buckwheat (*Fagopyrum esculentum*), ramie (*Boehmeria nivea*), rice, cotton, sorghum, jute (*Corchorus capsularis*), flax, grapes, opium poppy (*Papaver somniferum*), and trees such as gum (*Eucalyptus* species), citrus fruits, persimmon, and numerous forage crops had been established. Other crops, such as cinchona, coffee and tea had failed to take root and efforts to introduce them were discontinued in most parts of the country.[78] The current genetic landscape of the US agriculture started to take shape in the last century and has since then only undergone refinement. While new plants were introduced, established ones underwent major selection processes to produce the most suitable varieties, either by experimentation or global search, as the case of wheat illustrates.[79]

One major institutional change which affected seed distribution was the setting up of agricultural research stations under the Hatch Act of 1887. After the Agricultural Division became the Department of Agriculture, a large share of the seeds were sent to these stations and argicultural colleges. In 1893 the Department of Agriculture, under J. Sterling Morton, stressed that the introduction of rare flowers and trees be facilitated by the private sector, arguing that seed companies could get new varieties to the farmers three years ahead of the government. He called for the disbanding of the Seed Division, which had been set up in 1864, saying that its existence was an infringement on the rights of individuals engaged in private commercial activities. But Congressmen were unwilling to forgo the privilege of getting free seeds for their constituents.[80]

Additional institutional changes were introduced at the beginning of the 20th century. The divisions of Vegetable Physiology and Pathology, Agrostology, and Pomology were merged into the Office of Plant Industry. Together with the newly-formed section of Seed and Plant Introduction, the Office was renamed the Bureau of Plant Industry in 1901. By then work on plant introduction had become more systematic. The 'Inventory of Plants Introduced' had been started in 1898 by O. F. Cook, in the same year that the Office of Foreign Seed and Plant Introduction was set up. The inventory assigned a number to every new plant and provided information on its origin, nature, value and cultivation. It was also Cook who discontinued the designation of 'Special Agents' for agricultural explorers because the title raised suspicion. These agents travelled to foreign countries in search of economic plants and agricultural information.

The turn of the century was marked by increased botanic missions to various parts of the world. Most of the explorers went to Asia where they collected a large number of plants. Explorers such as Frank Meyer, P. H. Dorsett and Luther Burbank combed the world for new plants of economic value. By then the Department of Agriculture was shifting its focus from plant introduction to breeding. In the 1910s new varieties of wheat, barley, sorghum, cotton, dates, mangoes (*Mangifera* species), avocado (*Persea americana*), mangosteen (*Garcinia mangostana*), cassava (*Manihot esculenta*), dasheen (*Colocasia esculenta*), bamboo (*Bambusa* species), potato and soybean were added to the genetic landscape. By the 1920s the rediscovery of Mendel's work on genetics began to have a direct impact on agricultural production as breeders increasingly incorporated new genetic material in existing crops. By then plant introduction gardens disseminated plants that were suited to the locality and regional specialization had become the main feature of the US crop genetic landscape. Most of the major crops of the world had economic and agricultural niches somewhere in the US.

One of the most successful stories of crop substitution is that of soybean, a plant introduced largely to substitute for cotton in the South. Three varieties were imported from China, Japan and India in 1909 and additional varieties were later procured from Manchuria and Korea for extended cultivation northward. The Department of Agriculture later sent P. H. Dorsett and W. J. Morse to China, Manchuria, Japan and Sakhalin to collect the best varieties of soybean for US farmers. Over the 1929-31 period, they procured nearly 3,000 soybean varieties, in addition to other seeds such as alfalfa, barley, wheat, grasses, lespedeza (*Lespedeza* species) and mung beans (*Vigna radiata*).[81] A large number of economic plants was also added to the US collection by Frank N. Meyer, whose tenacity was ended in 1918 when he drowned in the Yangtze.

The need for self-sufficiency in major agricultural and industrial crops became more salient during World War II. These efforts were given a new boost in 1946 with the Research and Marketing Act which provided funds for federal-state co-operation in plant exploration, introduction and testing. Funds were also made available for cataloguing the existing genetic material and preserving it for future breeding activities. The Act allowed the Department of Agriculture to support more explorations.[82]

By the 1940s, US agriculture was already firmly based on a narrow range of genetic resources with strong emphasis on hybrids and monocultures, an agricultural model later exported to large sections of the world. It was a model that emphasized the superiority of high yields, a quantitative tradition that goes back to Bacon, Galileo, Descartes, and Newton. The world was combed for 'superior' genetic material that contributed to continued growth. Efforts to diversify agricultural production, especially in the South, were peripheral to the rise of monocultural production. The lessons of the potato rot epidemic in the 1840s were not effectively incorporated into the farming culture and plant breeding continued to depend on a narrow genetic base.

The research programmes emphasized genetic uniformity, and resulting varieties depended largely on the use of chemical inputs; the ecological uniformity of sections of the US tended to favour this agricultural model. Single varieties could be planted on a much larger scale than would be possible, say, in Africa and regions with more diverse ecological conditions. The rise of such an agricultural model was coupled with imperialistic (in the environmental and political sense of the word) aspirations of the US and the post-War period has seen the application of the Green Revolution in various countries of Asia and Latin America. While the model has increased the amount of food available to these countries, it has worsened social inequality and led to increasing environmental degradation. The promises of plenty have resulted in want.

This chapter has traced world historical botany from the early days of the Egyptian civilization to the rise of US agriculture. It has argued that the introduction of new genetic material into the agricultural system was a crucial innovation in economic development. The manner in which certain nations have pursued this evolutionary path has had profound effects on the economies and welfare of the Third World countries. The pioneering experiments of the US agriculture as well as the colonial efforts to transfer plants to various parts of the world have left an irreversible legacy, the effects of which are just starting to emerge. The following chapters examine this legacy and place it in its political and historical perspective.

Notes

1. See Reed (ed.), 1976, for details.

2. Some scholars, however, think it was the myrrh tree (*Commiphora myrrha*) that was taken to Egypt. Records also show plant hunting nearly 5,000 years ago when Sankhkara, one of the pharaohs, sent ships to the Gulf of Aden to collect cinnamon (*Cinnamonum verum*), and cassia (*Cinnamonum aromaticum*) used in embalming the dead. The history of Egyptian civilization is marked by persistent collection of plants from the region and their introduction in the Nile valley.

3. 'Two hundred and seventy-five [plants] are carved on the walls of the "Botanical Chamber" of his temple at Karnak, but only a few of them can be identified,' Coats, 1969, p. 243.

4. The Nile valley lies close to a major centre of genetic diversity. Highly sophisticated agricultural systems evolved in the region partly because of the availability of a wide range of genetic resources that could be utilized.

5. The monument reads: 'How magnificent is the result of Taji's work,' Ryerson, 1933, p. 110. By then, citrus fruits were already spreading to the West. The citrus spread started in prehistoric north-east India. By 136 BC, Simon the Maccabee was ordering that cedar cones be replaced by citrons in the Jewish harvest-festival ritual. Palestinian Greeks then conflated the two and kept the old name; all citrus plants are thus named after cedars (*kedros*). The citron (*Citrus medica*) was the first such fruit to reach Europe. This happened after Alexander the Great opened up good communication lines with the East. Columbus, on his second voyage of 1493, completed the global drift by transporting orange, lemon and citron seeds to Hispaniola. The pomelo (*Citrus grandis*) was introduced by one Captain Philip Shaddock in Barbados in 1696 but was never a popular fruit. The fruit later mutated to what is now called the grapefruit (*C. paradisi*), which reached the US in 1823 when a French settler, Odette Philippe, moved seeds or seedlings from the Bahamas to a site near Safety Habor on Tampa Bay. Most of the US grapefruit varieties originated from this introduction. Tangerines (*C. reticulata*) reached England from Morocco. From the original stock numerous hybrids have been introduced all over the world. For further details on the origin of the grapefruit, see Kumamoto, et al., 1987.

6. Their work was to undertake 'the methodical examination and identification of the products of nature of my American dominions, not only to promote the progress of the physical sciences, but also to banish doubts and falsifications which exist in medicine, painting, and other important arts, and to foster commerce, and to form herbaria and collections of the products of nature, describing and making drawings of the plants found in these, my fertile dominions, in order to enrich my Museum of Natural History and the Botanical Garden of the Court.' Steele, 1964.

7. Heiser, in *Economic Botany,* Vol. 40, No. 3, pp. 261-6.

8. See Brush, 1980.

9. See Baker, 1978, pp. 19-21.

10. The book was so useful that 'in 1903 a German scientist published a work of 400 pages on the botanical results of Alexander's expedition, as recorded by Theophrastus,' Coats, 1969, p. 143.

11. Lemmon, 1968, p. 2. 'So it was . . . that the Greeks took over from the Egyptians and so in turn it was that to a dismal British landscape of birch scrub, oaks, Scots pine and holly the Romans brought variety, charm, colour and usefulness . . . What a wealth of natural beauty and utility they brought! The plane tree, the lime, the chestnut, the sorbus, the box, the elm, the pear, the cherry, the vine, the damson, the quince, the peach, the mulberry, the fig, the rose, the medler, the hyacinth, rosemary, thyme, cabbage, leeks, onions, radishes, parsley and lettuce,'.

12. Lemmon, ibid, p. 3.

13. The sailing experience accumulated during the Crusades and the efforts to colonize the Azores and Canaries were instrumental in enabling Columbus to get to America and back.

14. Hobhouse, 1985, p. viii.

15. Not surprisingly the first known monopoly rights were issued to cooks in Venice to enable them to make the same dishes over a period. This is one of the roots of intellectual property protection.

16. Spices were so important that in some countries they became a unit of exchange or were used in place of money. For example, rents and taxes were often paid in the form of 'peppercorn' in medieval England.

17. See Rickett, 1956.

18. Brockway, 1979, p. 72. Botanic gardens also enabled botanists to undertake more systematic studies of plants. Methods of classification were suggested, tested, criticized and refined. This culminated in the publication of Linnaeus' *Genera Plantarum* in 1737. The growth of systematic botanical knowledge was therefore closely linked to the institutional organization of the day. Botanic gardens were not unique to Europe. The Aztec royalty maintained elaborate botanic gardens in Tenochtitlan, which were destroyed during the Spanish conquest. For details, see Lipp, 1976. The existence of such gardens in Mexico is consistent with the rise of complex uses of genetic resources in the region.

19. Brockway, 1979, p. 51.

20. In contrast, cocoa (*Theobroma cacao*) seeds were smuggled to Ghana by a labourer from São Tomé in 1878, and thus diffused among small-scale or peasant producers.

21. 'Not only were the crops improved and multiplied, but cattle, pigs, sheep, chickens, and turkey were bred into different shapes and sizes to yield more steaks, bacon, fleece, and larger eggs and drumsticks. In this atmosphere of heady scientific development, the landed gentry were more than ready to experiment with fruits, and vegetables in their own kitchen and pleasure gardens,' Morwood, 1974, pp. 11-12.

22. Quoted in ibid, p. 12.

23. Douglas, possibly the greatest plant hunter of all times, is popularly remembered for his introduction of the Douglas fir (*Pseudotsuga taxifolia*). But his immense work and contribution to the British gardens is immortalized in numerous plants which bear his name. In Munz and Keck's *A California Flora,* for example, Douglas's name (as *Douglasiana* or *Douglasii*) is attached to 86 species or varieties.

24. The two plants were later identified as annual bluegrass (*Poa annua*) and *Nephrodium* (now *Dryopteris*) *filix-mas*.

25. The role of technological innovation in revolutionizing plant transportation and therefore transforming the plant genetic map of the world is often underestimated. Klose, for example, accords this significant discovery a mere footnote: 'The use of Wardian glass cases for transporting plants great distances at sea came to be widely practised soon after the discovery of their principle by a London physician, Nathaniel B. Ward in 1829. The Wardian case is simply a closed glass case that protects plants from various unfavourable conditions. It protects them from impure air, salt spray, cold air, and high winds. It maintains constant humidity and moisture in the soil, because it permits only negligible air circulation. With the advent of transportation by airplane, the Wardian cases have become largely obsolete', Klose, 1950, p. 47. Given the limited crop genetic resources in North America, the rise of a sophisticated agricultural economy could have been difficult without the help of the Wardian case.

26. This view is stated clearly by Schoenermarck: 'Like many marvellous and useful discoveries (which we often tend to mislabel inventions), the fact that plants could be grown in closed containers – today called terrariums – was found out quite by accident,' Schoenermarck, 1974, p. 148. This is repeated by Brockway, 1979, p. 87.

27. 'Apprenticed to his father's profession, he studied at the London Hospital and attended "botanical demonstrations and herborisings" of Thomas Wheeler, then demonstrator to the Society of Apothecaries. (We do tend to forget in these days of ersatz pills, that, until their advent, all medicines were compounded from plants rather than coal-tar and the like). What better way then, for Dr. Ward to combine his first love, botany, with

his profession, medicine,' Schoenermarck, 1974 p. 149.

28. 'Dr. Ward had long tried and failed to grow ferns and mosses on, and at the bottom of, an ugly, dirty, sooty old wall facing his surgery. He never did make anything grow on it. But the succes of his enclosed cases opened up a vast field of horticultural adventure to him, and more than made up for previous failure in the open, smoky air of London', ibid., p. 150.

29. Lemmon, 1968, p. 183.

30. Schoenermarck, 1974, p. 150.

31. The shipment included *Gleichenia microphylla* and *Callicoma serratifolia*.

32. 'It was February, 1834, in Sydney, with the thermometer hovering between 900 and 1000, when the cases were lashed to the decks of the homeward-bound clipper. While the ship was rounding Cape Horn, the temperature dropped to 200 and the decks were covered by more than a foot of snow. While crossing the Equator, the thermometer rose to 1200. When the ship entered the English Channel in November, some eight months after leaving Sidney, the temperature was only 400. The Plants had received no protection whatsoever, either by day or by night, were never once watered, yet when they were taken out . . . they were in "the most healthy and vigorous condition," ' Schoenermarck, 1974, p. 152.

33. The successful introduction of *Amherstia nobilis* too established the viability of Wardian cases as a major innovation in plant transportation. Among other plants, Gibson introduced orchids such as *Coelogyne gardneriana*, *Dendrobium devonianum*, *D. gibsonii*, and *Thunia alba*. He also introduced *Rhododendron formosum*. 'Orchidomania broke out, and for most of the remainder of the century orchid hunters were everywhere tearing these fantastic beauties from their natural haunts in their thousands. Whole tracts of forests were mown to the ground to get the epiphytal treasure. This over, then much of the lure of the tropical, subtropical and antipodean exotics had gone, mainly – and this is no exaggeration – because most of the desirable, transferable and growable plants had been collected. Higher, hardier plants were now to be sought for to suit a new fashion in gardens and gardening, the William Robinson era, of hardy herbaceous and rockery plants, as the Farrers, Wilsons and Wards of this world travelled the mountainous Asiatic ranges', Lemmon, 1968, p. 218.

34. Ward's contribution to horticultural development was rewarded when Willian J. Hooker and Willian H. Harvey named a genus of South African mosses *Wardia*. He was recognized by his peers both in medicine and botany. He was the Examiner for the Society of Apothecaries for nearly 20 years, and became Master of the Society in 1856 and later its Treasurer. In 1852, at the age of 26, he was elected a Fellow of both the Linnaean Society and the Royal Society.

35. The cases attracted much attention from the scientific community. Michael Faraday lectured on the theme of 'Growing Plants Without Open Exposure to Air', at the Royal Institution in 1838.

36. This was not the first time that a European country turned to China for sources of innovations. For centuries China was a rich source of technological innovations not widely used in economic activities; gunpowder, magnetism, paper and printing, the wheelbarrow as well as numerous metal-working devices, were among innovations transmitted from China to Europe.

37. See Cox, 1945, pp. 34-69, for a more comprehensive account of the commercial links between the East India Company and plant collection in China.

38. 'The flora of China might have been even more rich, had not its ancient civilization and large population entailed the destruction of all wild plants by intense cultivation over vast areas, long before the first plant collectors arrived. Probably, however, the cultivable plains were never so rich in species as the wild mountain areas, which remained virtually untouched until the late nineteenth century', Coats, 1969, p. 87.

39. Tea was first directly shipped to Britain in 1669, previously Britain received tea through Dutch re-export facilities in Indonesia. For centuries the country spent foreign exchange on a plant whose origin was not even known.

40. Fay, 1985, p. 18.

41. 'Early in 1820 David Scott, commissioner for the newly acquired state of Assam . . . sent samples of leaves . . . to his superior in Calcutta. Here they were declared to be one of the . . . species of camellia, and sent by Dr Wallich, the government botanist, to London where they were re-examined by the herbalist of the Linnean Society, and pronounced to be from the tea plant. This was the first known identification of a wild tea plant in India, which at the date had no tea plantations. At this time nearly all tea came from China, except for a small export surplus available in Japan, and even less in Formosa,' Hobhouse, 1985, pp. 123-4.

42. The tea genus, *Camellia*, was named after Came (Latinized as Camellus), a Moravian Jesuit (1660-1706) who wrote extensively about Asian plants. In his *Genera Plantarum*, Linnaeus put all teas under the name *Thea sinensis* and noted two camellias, *Camellia japonica* and *C. sassanqua*. In 1762 Linnaeus distinguished two tea varieties, *Thea viridis* and *T. bohea* (for 'green' and 'black' tea). However, Camellia and Thea are members of the same genus, *Camellia*, which belongs to the Theaceae family (of 240 species of which only two are of major economic value).

43. 'Many botanical gardens in the tropics subsequently fell on hard times. Some, such as the one established by the French at Port au Prince, Haiti, closed', Plucknett and Smith, 1986a, p. 306.

44. The genus was named after the wife of the Viceroy of Lima, Luis Fernandez de Cabrera Bobadilla y Mendoza, the fourth Count of Chinchon, who was cured of malaria by an extract from the plant. Linnaeus misspelt the plant as 'Cinchon' in his *Genera Plantarum*. The transfer was justified on the grounds that the plant was facing extinction due to excessive cutting and wasteful utilization by the local community. This view prevailed despite the

fact that previous studies had shown that the plant was better managed through coppicing and not the mere removal of its bark. Coppicing allowed for new shoots to grow and be ready for harvesting in a few years while the removal of the bark made the plant susceptible to pests.

45. Brockway, 1979, p. 114. The collection was made on land belonging to the Catholic Church and a former President of Equador engaged in anti-government activities. While Spruce and Cross collected the 'Red bark', Clements Markham collected the 'yellow bark' and the 'grey bark' in Bolivia and Peru. Not much came out of his collection.

46. The official link between the India Office and Kew ended in 1862 and all the subsequent collections went to Kew. Kew, however, continued its interest in cinchona development through its representatives in India.

47. The plant also spread to other countries. Farmers in Ceylon turned to cinchona after leaf blight destroyed their coffee in the 1870s. Later the Ceylonese farmers switched to tea as the red bark started losing the market to the Dutch Ledger bark which was easier to process into sulphate of quinine and could therefore be easily regulated.

48. Corporate motives have not changed much since the days of cinchona development last century. Recent debates at the World Health Organization (WHO) have dealt precisely with the same issue regarding the choice of vaccines against malaria. The corporate logic favours the more expensive alternatives, which would be inaccessible to most of the world's poor people.

49. The relocation of plants was in some cases associated with the relocation of labour. Once new plants were available, it was necessary to bring in people to work on the farms. This part of the history of economic development is well documented with various interpretations; slave trade is the classical case. Additional details on the relocation of plants such as rubber and sisal can be found in Brockway, 1979, Chapters 7 and 8.

50. Australia used even less of its indigenous plants in modern agriculture, but relied entirely on exotic material. See Parbery, 1964, for a report on the role of the South-East Asia Treaty Organization (SEATO) in plant introduction in Asia and Australia in the early 1960s.

51. Although corn is a major crop in the US agriculture, its origin is Mexico and Central America. By the time European settlers arrived in North America, the Indians had already introduced corn and 'Irish' potatoes. The settlers sent numerous plants that had been introduced by Indians to Europe for trial and reintroduction there including the agave, arrowroot, kidney beans (*Phaseolus vulgaris*) and lima beans (*Phaseolus lunatus*), cacao, chili pepper (*Capsicum annuum*), cashew nuts (*Anacardium occidentale*), cherimoya (*Annona chrimola*), cotton, gourds (*Cucurbita* species), guava (*Psidium guajava*), Jerusalem artichoke, cassava, maté (*Ilex paraguariensis*), papaya (*Carica papaya*), peanut (*Arachis hypogaea*), pineapple (*Ananas comosus*), prickly pear (*Opuntia vulgaris*), pumpkin (*Cucurbita* species),

quinoa (*Chenopodium quinoa*), squash (*Cucurbita* species), sweet potato (*Ipomoea batatas*), tobacco, and tomato (*Lycopersicon lycopersicum*).

52. 'If pathogens could pass freely from the Old World to the New World, so . . . could other life forms. So could the life forms that provide man with food, fiber, hides, and labor, that is, cultivated plants and domesticated animals. To a notable extent, the whole migration of Spaniards, Portuguese, and the others . . . and [their] successful exploitation of the New World . . depended on their ability to ''Europeanize'' the flora and fauna of the New World. The Transformation was well under way by 1500, and it was irrevocable in both North and South America by 1550', Crosby, 1972, p. 64.

53. Sugar-cane was introduced in Europe by the Crusaders. For a fuller account of the story of sugar, see Mintz, 1985.

54. Of these ingredients, olive oil was the most difficult to produce in the colonies. Early Spanish settlers recognized that olive trees could grow in the drier valleys of Peru and Chile; but the seedlings had to be brought in from Europe. Not until 1560, long after wheat and vines were established in Peru, did Antonio de Rivera bring seedlings from Spain. 'Only two or three . . . had survived the journey. Their value was so immense that he posted a number of slaves and dogs to guard them. It was no use: one was stolen and spirited off some five hundred leagues south of Chile. These seedlings, whether legitimately or illegitimately acquired, were the beginning of what quickly became a considerable olive oil industry in the irrigated valleys of South America's arid Pacific coast', Crosby, 1972, p. 73.

55. Ibid, p. 75.

56. In 1540 Gonzalo Pizzaro collected horses, llamas, dogs and 2,000 pigs for an expedition to the east of the Andes to look for the Land of Cinnamon.

57. Crosby, 1972, p. 66.

58. Jefferson, T., quoted in Brockway, 1979, p. 35.

59. He fully supported the activities of the South Carolina Society for the Promotion of Agriculture which was established in 1785 and was the first association incorporated in the US to provide a farm for testing imported plants. Most of the early work was on testing the grape and olive. Jefferson was an active member of the Albermarle Agricultural Society of Virginia, founded in 1817 by his county neighbours.

60. The people saw public sector institutions as crucial in the development of agriculture. 'In New York, *The New York Genessee Farmer* of 1836 praised farmers of the Monroe County for presenting a petition to the New York legislature for an appropriation to aid a state agricultural institution at the head of the county agricultural societies. The money was to be spent for premiums for agricultural products and for procuring useful seeds for public distribution', Klose, 1950, p. 20.

61. The local people were aware of the implications of letting economic plants fall in the hands of potential competitors so they made it difficult for

Perrine to have access to seeds. At times he had to provide medical services in exchange for seeds.

62. Perrine was shot in a Seminole uprising in 1839. His plants, which included 200 varieties of tropical plants, were destroyed in the attack. Some of the remaining plants were transferred to greenhouses in the north or taken by army officers as ornamentals.

63. The circular read in part: 'The introduction of useful plants, not before cultivation, or of such as are of superior quality to those which have been previously introduced, is an object of great importance to every civilized state, but more particularly to one recently organized, in which the progress of improvements of every kind, has not to contend with ancient and deep rooted prejudices. The introduction of such inventions, the results of the labour and science of other nations, is still more important, especially to the United States, whose institutions secure to the importer no exclusive advantage from their introduction . . . The collectors of the different ports of the United States will cheerfully cooperate with you in this interesting and beneficial undertaking, and become the distributors of the collections of plants and seeds which may be consigned by you to their care. It will greatly facilitate the distribution, if the article shall be sent directly to those sections of the Union, where the soil and climate are adapted to their culture. At present, no expense can be authorized, in relation to these objects. Should the result of these suggestions answer my expectations, it is possible that the attention of the national legislature may be attracted to the subject, and that some provisions may be made, especially in relation to useful inventions', quoted in Klose, 1950, p. 26.

64. This was the origin of the propagating and botanic gardens which were expanded 25 years later as the demand for new plants increased.

65. In 1827, the Navy Department ordered Commander William Crane thus: 'It will probably be in your power, while protecting the commercial, to add something to the agricultural interests of the nation, by procuring information respecting valuable animals, seeds, and plants, and importing such as you can conveniently, without inattentions to your more appropriate duties, or expense to the Government. There are many scientific, agricultural and Botanical institutions, to which your collections might be profitably instructed, and by which whatever you procure will be used to the most extensive advantage of the country. Among those is the Columbian Institute of this city', quoted in Klose, 1950, p. 28.

66. Other naval officers brought seeds and distributed them among friends. For example, one Captain Ballard brought five bushels of *lupinella* from Italy and distributed them among his friends near Annapolis.

67. The instructions also said: 'You will keep a full and accurate Journal of proceedings and operations, which on your return, will be delivered by you to this Department', quoted in Klose, 1950, p. 146.

68. In 1855 Congress allocated US$1,500 for a building to house the

Japanese plants. At the time Perry was in Japan, another naval vessel, the *Water Witch,* was exploring the Paraguay River. The seeds collected on this mission, which included maté (Paraguay Tea) were sent to the Patent Office.

69. The Patent Office argued that the increased yield from the *Release* introduction more than compensated for the damage caused by the cane borer.

70. Most of the 200 or more varieties grown in the US in the 1950s descended from the consignment of an unknown diplomat in Panama. The man who bred the new varieties, Rev. Chauncey Goodrich, did not know where they came from. He called two of his new varieties 'Rough Purple Chili' and 'Garnet Chili' assuming that they came from Chile.

71. The South American Commission of 1817-18 was sent to the region to monitor political events but William Baldwin, a member of the team, made valuable plant collections. '[His] collection of plant specimens was available for study in the Academy of Natural Sciences of Philadelphia, [his] notes were of assistance to contemporary botanists, and his work aroused scientific interest in South American plants', Rasmussen, 1955, p. 31.

72. Innovations and new genetic information were considered as endogenous to economic growth. But this was before the Newtonian notions of equilibrium were applied to economic theory and public policy, thereby allowing innovations and genetic material to be treated as exogenous to economic evolution. These notions are discussed in Chapter 1 and in more detail in Clark and Juma, 1987. See also Juma, 1986.

73. Similar arrangements were made with foreign botanic gardens and societies such as Kew, Melbourne, India Museum in London, Cape of Good Hope Agricultural Society, British Museum, Central Agronomical Society of the Grand Duchy of Posen, Horticultural Union Society of Berlin, Royal Society of Brussels, Royal Gardens of Madrid, Horticultural Society of Bremen, Royal Meteorological Society of Edinburgh, and the Agricultural Society of Sydney, New South Wales. Seeds and plants were also freely sent to the US from the Imperial Gardens at Tokyo, especially after the Perry expedition.

74. One of his main introductions was sorgo.

75. Foreign ministers, missionaries, consuls, merchants, travellers, and naval officers were requested to collect seeds in foreign countries and send them to the US. The postal franking privileges facilitated the wide distribution of seeds.

76. Over the same period, forest species of economic and ornamental value were also being collected and distributed. The distribution of seeds of the Japanese camphor tree started as early as 1862; osier willow seeds (*Salix viminalis*) were imported from Europe and widely distributed for making baskets and rough furniture. Equally important was the cork oak, introduced for the wine industry. Cork production, however, remained sluggish because it required abundant cheap labour. For timber, seeds of the Chinese laurel tree

(*Laurus nobilis*) sent from the US consulate in Shanghai were widely distributed. Other tree introductions included the Japanese mulberry, Chinese camphor, eucalyptus and the English oak (*Quercus robur*).

77. See Ross, 1946, for details on this tension.

78. In its efforts to introduce tea, in 1858 the US enlisted the services of Robert Fortune, who, a decade earlier, enabled the British to transfer tea from China to India. After his first consignment of tea seeds and plants aboard the *Nabob,* the Commissioner of the Patent Officer, Joseph Holt, decided to terminate Fortune's contract before he returned to the US. Holt felt that after receiving the seeds and plants, it was no longer necessary to retain an expert. In 1881 Congress provided US$15,000 for establishing a tea experimental farm at Summerville, South Carolina.

79. See Ball, 1930, for details.

80. They responded by increasing the appropriation for seed distribution to US$130,000 in 1895. The following year saw record seed distribution; Morton was incensed, He devoted 79 pages of the Department's Annual Report to a list of addresses of the recipients of seeds under the congressional franking privilege.

81. For a detailed account of their contribution to the US soybean industry, see Hymowitz, 1984. The soybean collection also benefited from the work of C. V. Piper.

82. While Jack Harlan was sent to Turkey to collect vegetables, forage plants, oil crops and cereals for breeding, W. Kelz went to India to search for winter forage crops and potential industrial plants, and D. S. Correll brought home disease-resistant potato varieties from Mexico. The expeditions enriched the US stock by some 4,500 introductions over 1946-49 alone.

3. Genetic Resources and World Agriculture

Plant collection formed the early stages of the internationalization of agricultural research. By the end of the 19th century, most innovations arising from the Western countries were already being applied in other continents. Japan, for example, introduced agricultural policies which encouraged the importation of Western technology. By the 1940s the US had developed a monocultural agricultural model that relied on a narrow range of genetic material. It was guided largely by reductionist logic and had at its disposal a large stock of technological innovations. The stage was already set for major global changes arising from these innovations. The process, it appears from available evidence, moved from national to regional impacts.

But as the various world nations became gradually interlinked, it became possible to undertake major agricultural research programmes which affected large sections of the world. The programme was also associated with the complex organization of the flow of genetic resources to meet the requirements of agricultural production. This is exemplified by the Green Revolution in Latin America and Asia and subsequent efforts to promote the model in other parts of the world. The Green Revolution had both gains and losses. At its core was the expanded use and flow of genetic resources;[1] this use was associated with the complex evolution of institutional networks and corporations which maintained a firm grip on key genetic resources used in global agriculture.

Japanese antecedents

In order to understand the historical evolution of the Green Revolution and its current ramifications, it is important to locate the process in a long-term perspective. The Green Revolution usually refers to the use of high-response varieties (HRVs) of wheat and rice in Latin America and Asia respectively. What is often forgotten is that the introduction of improved varieties was undertaken by the Japanese long before the methods were applied in Mexico and Asia. Indeed, it is no surprise that both the Mexican and Indian revolutions were possible partly because of the incorporation of genetic material from Japanese breeding efforts in wheat and rice. Today, Japan is often featured as an industrial giant but its contribution to international agriculture is often ignored.

The innovative capability in Japanese agriculture can be traced back to the Tokugawa period (1600-1868) when the country was in isolation. This period was marked by the introduction of a wide range of innovations, but the prevailing feudal restrictions on travel and communication limited their widespread application; the Meiji Restoration allowed for their extensive adoption. Moreover, the innovative capability that had been developed over the centuries enabled Japan to rapidly adopt imported technologies and combine them with existing productive facilities.[2] The agricultural history of Japan is also associated with the contribution of their innovations to other countries: the dwarfing genes used in the Green Revolution wheat varieties reached the US in 1946 through the Japanese Norin 10 wheat. These genes later contributed greatly to US and Third World wheat output.

Japanese research in local varietal improvement was intensified in the 19th century to provide staples for the urban population. By then, European powers were increasing production in the colonies. Japanese peasants played a major role in the selection of high-yielding varieties. The country made efforts to import Western technology, mainly from the US and Britain. Attempts to introduce large-scale farm machinery were abandoned in 1888 when the Mita Farm Machinery Manufacturing Plant was sold by the government to the private sector.[3] Because of the prevalence of small farms in Japan, the country turned to intensified agriculture as a way of increasing yields. Japan forged closer links with Germany in order to acquire chemical technology for the agricultural sector. By the 1890s Japan experienced rapid urbanization which was associated with increased industrial production. Since the country could not meet all the food needs for the population, efforts were undertaken to search for alternative sources. One solution was to use colonies as a source of food. Taiwan (then Formosa), which was occupied by the Japanese in 1895, supplied the country with sugar but its local rice varieties did not appeal to the Japanese; it could not, therefore, be marketed in Japan.

Reasoning by analogy, Japanese administrators sought to replace the local rice varieties with those that could be sold on the Japanese market. Research into ways of introducing new varieties, initiated soon after the occupation, was done systematically and relied on a long-term agricultural strategy. A 10-year survey of all the rice grown on the island was undertaken, the results of which showed that in 1915 there were some 1,197 native varieties under cultivation. Attempts to improve the quality and yield of the varieties began in 1909 and in 1910 a 'Native Improvement Programme' was started.[4]

This programme had three main objectives. First, it sought to reduce the number of poor-quality and low-yield varieties and to encourage the farmers to stop growing a wide range of varieties and concentrate on selected seed. Second, the programme aimed at eliminating the red rice, a sturdy drought-resistant variety that was not palatable to the Japanese but was widely grown by Taiwanese farmers. Third, the programme undertook research to select a variety from the local genetic pool which was stable, less chalky and easier to dry; it also needed to be soft enough to meet Japanese consumer preferences. The police conducted extensive campaigns against local varieties, especially red rice, and so effective was the process that in the decade after the launching of the Native Improvement Programme, the number of cultivated varieties had been reduced to 390, a cut of 67 per cent.

One of the main features of the Japanese breeding programme was the introduction of varieties that were responsive to fertilizers. 'The use of artificial fertilizer increased by 254 per cent, the average rice yield increased by almost 50 per cent and an average increase of 4.3 per cent per annum was achieved in gross farm output, in which improvement in the yields of sugar played an important part.'[5] This increased the sale of fertilizers and therefore boosted industrial output in Japan, a feature that subsequently became a major characteristic of the Green Revolution. It also became one of the main sources of criticism, for it not only shifted production into the hands of rich farmers who could afford the fertilizers, but also led to environmental pollution.

In addition to varietal selection, the Japanese introduced irrigation, deep ploughing, artificial fertilizers, intensive planting, green manure and pest control measures to improve rice productivity. The efforts, however, were not successful and overall total farm output increased by only 1.7 per cent per year over the 1906-20 period. Increasing productivity depended on introducing new varieties. In 1922, plant breeders succeeded in crossing *japonica* and *chailai* varieties and achieving marketability, fertilizer responsiveness and quick maturation. The new variety, *ponloi*, was widely planted in Taiwan and by 1922 covered 400 hectares and by 1940 the coverage had increased to 324,000 hectares, more than half the country's rice crop.

The Japanese experience influenced subsequent efforts to introduce HRVs in Asia, especially after World War II. Institutions such as the International Rice Research Institute (IRRI) owe much to these historical antecedents. The Taiwanese case study raises a number of important issues. In the first place, imperatives to raise production in specific varieties for the Japanese market required a parallel programme to reduce the prevalence of genetic diversity in rice. This was achieved through administrative and coercive measures, but the very introduction of HRVs was bound to displace the local varieties.

The Green Revolution was an analogue of the mass production paradigm in industry. The underlying philosophy and tools applied to the introduction of HRVs resemble industrial production in various ways. Mass production relies on the 'use of special-purpose (product-specific) machines and of semiskilled workers to produce standardized goods'.[6] Mass production requires the existence of a combination of standardized components and procedures that make output responsive to economies of scale. While such parts could be produced through research and development (R&D), finding their genetic equivalent required a long period of search and breeding. After the introduction of a new variety suitable for particular political, economic and social objectives, the Green Revolution approach focused on increasing output. Like its industrial analogue, the Green Revolution was associated with the concentration of resources and power among a smaller number of farms. Its expansionist tendencies are explicit in the use of a few crop varieties under controlled conditions over a large area.

The adaptive knowledge that had hitherto been accumulated at the local level for dealing with a wide range of agricultural problems such as pests, drought and disease was scrapped, and improved varieties came with new techniques of how to deal with the problem. This is similar to the way craft knowledge was replaced by factory production which required mundane and narrowly-defined skills. The Green Revolution not only made the farmers dependent on the suppliers of seed and inputs, but it also made them less adaptive as they had no control over the knowledge related to the new varieties and inputs. Like in mass production, production knowledge was moved from the farm and located in research institutes from where it was entrusted to the hands of extension officers. The farmers thus lost part of their autonomy and the capacity to conduct adaptive experiments. Whenever the new varieties failed, the farmer was left with very few options but to await solutions from the research centres. Control over the production process was thus removed from the farmers.

In addition to these issues, the Japanese rice programme was not designed to benefit the Taiwanese people but to meet the food needs of the Japanese empire. The research agenda was therefore defined by the economic, political and dietary requirements of the Japanese market. The social and political shaping of scientific research, and the resulting technologies, has continued to be a major factor in breeding programmes. The Japanese example in Taiwan was later replicated and widely imposed on a number of Third World countries by international institutions in collaboration with national governments. Local knowledge was denigrated. The widespread application of the model was associated with the growth of the global seed industry.

Plant breeding and the seed industry

The seed industry has varied origins governed largely by the social and economic factors influencing agricultural production. In Europe, for example, the industry was associated with the changing patterns of land ownership as well as the rising complexity of agricultural production and the related accumulation of knowledge on plant development. The landed aristocracy and the church had access to land that could be used for breeding purposes. As agriculture became more complex, activities such as grain storage and shipping diversified into seed cleaning and trading.[7] Seed distribution in the US was facilitated by members of the Shaker religious community who were the first to start packaging seeds.

The rise of the modern seed industry is closely associated with plant breeding, an activity that has a long history, and it is difficult to pinpoint its landmarks. Major advances in plant breeding, however, were made after the rediscovery of Mendel's work on genetics. Knowledge of genetics made it possible to undertake more controlled breeding programmes and therefore reduce the amount of time and material used in the exercise. One of the most significant developments in plant breeding which helped reorganize and strengthen the seed industry was the production of hybrids.[8] Hybrids not only increase farm output, but also guarantee effective seed supply control by the breeders.

Hybridization reduces the viability of seeds over generations or progenies. This ensures that the farmer will return to the breeder every year for seed and therefore maintain the turnover of the industry. In addition to the production of hybrids, the industry has been shifting towards seedless crops, especially fruits and vegetables. This breeding programme affords the industry three main advantages. In the first place, seedless fruits ensure that the breeders have control over the reproductive material; secondly, reduction in seed content allows more room for the commercially useful part of the fruit or vegetable; thirdly seedless fruits or vegetables may be more suitable for industrial processing because of uniformity in texture.

Although hybrid crops enabled the breeders to have control over germplasm, they could not be patented. As a result, it was possible for competitors to develop similar varieties and therefore reduce the potential for monopoly profits. To avoid this, the hybrid industry sought to reduce the potential for the release of competing varieties by putting pressure on public sector institutions to stop work on hybrids. By the early 1950s, the US public sector institutions stopped producing hybrid corn; the pressure was intensified and the public sector has continued to withdraw from the release of other hybrid varieties since then.[9]

The use of hybrid varieties helped rejuvenate the US seed industry, which experienced slow growth rates early this century. With anticipated growth potential, especially with the use of hybrids, the industry began to expand although it remained diverse in the 1930s. This period, especially after the Plant Patent Act of 1930, led to increased private breeding activities. The struggle against public sector breeding that started in the previous century continued, and the Act provided a legal basis within which to continue the struggle. By the 1960s 10 major national US firms were selling seeds, with numerous others selling to various parts of the world.[10] By then, the seed industry was engaged in a major campaign to introduce an international system that would give patent-like protection to new plant varieties. The signing of the Union for the Protection of New Varieties of Plants (UPOV) Convention in 1961 provided renewed impetus for private breeding activities.[11]

During the 1960s, the seed industry remained relatively sparse although seed sales increased, partly as a result of the demand for the Green Revolution. Not until the early 1970s did the seed industry begin to experience major restructuring, as a result of the Plant Variety Protection Act (PVPA) of 1970 which provided breeders with wider patent-like rights. The 1970s saw major changes in commodity prices, especially grains. These increases raised new awareness of the potential profitability of the seed industry, and this perception, coupled with the property rights granted by the PVPA, contributed to a wave of corporate mergers and takeovers.

The main trend in the 1970s was that of large agro-chemical firms taking over seed companies. By doing so, the agro-chemical companies could now benefit from commercial as well as R&D synergy between seeds, agricultural inputs and marketing. It became possible for the firms to sell seeds together with other agricultural inputs in a package. The prospects for advances in biotechnology, especially after 1974, also stimulated the takeovers and the related concentration of the industry. By then, the agro-chemical firms had already accumulated extensive expertise in R&D which could be turned to biotechnology research. Moreover, the marketing channels developed by the seed companies over the decades could now become a conduit for biotechnology-based products.[12]

In addition to these factors, it is notable that the chemicals industry was itself undergoing a major slump in innovation. The US chemicals industry had experienced rapid growth over the 1950-74 period, nearly twice that of the gross national product (GNP).[13] Much of the growth was accounted for by the substitution of natural products, which peaked in 1974, when synthetic fibres accounted for 70 per cent of the total fibre market. This growth was stalled by improvements in treating cotton and wool as well as consumer preference for natural products. The industry was experiencing a slump in innovation and therefore facing industrial stagnation. From the 1930s to the early 1980s, there were 63 major innovations in the industry of which 40 were introduced in the 1930s and 1940s, 20 more in the 1950s and 1960s and only three in the 1970s and 1980s.[14]

Table 3.1
The world's leading seed firms, 1986 (US$ million)

Parent Firm	Industry	Seed Sales	% of Group Sales	Seed Firms	% of World Market
Pioneer Hi-Bred (US)	Seed	734.5	89.4	38	4.1
Sandoz (Swiss)	Chemicals	289.8	8.0	36	3.2
Dekalb-Pfizer (US)	Petrochemical	201.4	40.0	34	2.2
Upjohn (US)	Chemicals	200.0	10.1	15	2.2
Limagrain (French)	Seed	171.6	85.0	22	1.9
Shell (Anglo-Dutch)	Petrochemical	350.0	0.2	70	1.9
ICI (UK)*	Chemicals	160.0	1.1	—	1.9
Ciba-Geigy (Swiss)	Chemicals	152.0	2.0	31	1.7
Orsan (French)	Chemicals	119.0	53.0	—	1.3
Cargill (US)	Agribusiness	115.0	0.5	29	1.3

* Includes 1986 acquisitions
Source: Robert Fleming Securities, quoted in *The Economist*, 'Fruit Machines'; Mooney and Fowler, *Community Seed Bank Kit*.

The industry also faced other major problems in the 1970s. While the price of raw materials (especially petroleum) rose, the fluctuations in the exchange rate made international investments problematic. The rise of the environmental movement in the US and increased legislation against the use of a wide range of chemicals forced the industry to look into alternative innovations or investment strategies. Over the 1982-83 period the industry experienced its severest recession in 50 years. Major US companies that had invested heavily in Europe started to retreat and relocate some of their production facilities in other countries where raw materials were cheap and environmental regulations less stringent. In addition, the firms started diversifying their R&D activities into speciality products as well as pharmaceuticals. The increased participation of chemicals firms in the seed industry is part of a larger strategy to restructure their operations and cast a wider net for new sources of industrial renewal. The process did not only include acquisitions and mergers but it also involved divestitures. In 1986 alone there were 26 divestitures valued at over US$100 million each. However, the trend has been towards the control of most of the major R&D operations by fewer and stronger firms.[15] This concentration has also affected the seed industry.

The seed industry as a distinct entity is being eroded and the operations are being undertaken in conjunction with other activities under larger and broader investment strategies. The categories of major firms in the industry range from agribusiness companies to automobile producers. As early as 1974 Sandoz entered the seed business with the takeover of Northrup King (USA). This was followed by the acquisition of other seed companies such as Rogers Seed (USA) and Zaadunie, Sluis and Groot and Caillard in Holland and France. Sandoz's seed sales are now comparable to their income from plant protection chemicals.

Ciba Geigy entered the field with the acquisition of Funk Seed International (USA) and started producing hybrid corn and sorghum. The firm has an international network of 26 breeding stations all over the world and has built a molecular laboratory in North Carolina (USA) to complement its work at Friedrich Miescher Institute at Basle (Switzerland). In addition to this, Ciba Geigy has enhanced its plant tissue culture capability by buying shares in Phytogen (USA) and acquiring Hartman's Plants (USA) as well as signing a contract with the Agricultural Genetic Corporation, a British public sector research body.

Royal Dutch Shell took over the Nickerson company in 1978 and has continued to acquire seed firms since then. Nickerson is operating in over 70 countries and is one of the largest seed firms in the world. It was the first firm to market hybrid wheat, which was developed with the help of an agent produced by Shell's research centre at Sittingbourne, UK. In its efforts to diversify its activities, Monsanto has acquired a large number of seed firms in recent years. The firm now controls the operations of Jacob Hartz Seed, Hybritech Seed International, Monsanto Seed, Farmers Hybrid and numerous other firms. Its interest in biotechnology is extended to contracts with universities and it has agricultural research stations in the US, Japan, Brazil and Belgium.[16]

Recent major changes in the seed industry include the entry of the Imperial Chemical Industries (ICI) in the business. In 1985 ICI acquired Garst Seed (USA) for UK£60 million and later Sinclair McGill (UK) for UK£5 million. Garst is the second largest maize producer in the US and Sinclair McGill is a leading supplier of cereal seed in the UK. In 1987, ICI paid US$150 million for Belgium's Société Européenne de Semences and planned to sell UK£200 million worth of seeds in 1988 and double that amount by 1999. ICI has also increased its activities in the pesticide field despite the fact that sales in this industry have been sluggish. In 1987, ICI bought the US pesticide firm, Stauffer. The corporate strategy of ICI is to move into herbicide-resistant and pesticide-responsive crops which will help the firm increase its agrochemical sales.[17]

Amid the restructuring, government institutions are also being taken over by private firms under the privatization philosophy prevailing in the West. One of the most significant developments has been the buying of Britain's prestigious state-owned Plant Breeding Institute (PBI) at

Cambridge and the National Seed Development Organization (NSDO) by Unilever in 1987.[18] The deal cost Unilever UK£66 million. By buying PBI, Unilever now has control over research on new crop varieties that are resistant to disease and can undermine the sales of pesticide. PBI is active in this field and has already developed a new wheat variety, Rendezvous, which is resistant to eyespot, a fungus that attacks the base of the plant's stem.[19]

The current restructuring of the seed industry is associated with diversity among the key firms. The growing significance of seeds is illustrated by the entry of the Swedish automobile firm, Volvo, into the industry with the control of 47 seed companies. Given the uncertain nature of innovations in biotechnology, the reorganization of the industry is expected to continue well into the next century. The changing composition of the seed industry, in addition to the emerging biotechnologies, will introduce major changes in world agriculture. The potential for the industrial production of new seeds is expanding and the Third World countries are likely to be major destinations for these new seeds. Public sector breeding programmes in the Third World are likely to be affected by the current trends in the seed industry. Not only are these programmes subject to government policy changes, but some of them rely on external sources of funds for some of their activities and can, therefore, be easily influenced.

These changes are likely to affect the consumers in various ways. In the first place, the control of the seed industry by a few large firms may tend to keep prices at levels that can only be afforded by rich farmers despite the fact that biotechnology opens up opportunities to breed for the poor farmers as well. It is not necessarily the concentration of the industry that will lead to price increases but the R&D investment that is going into the production of new varieties. Some of the costs of the current acquisition activities may subsequently be passed on to the consumers. It is estimated that biotechnology will add value to seeds that will raise the price by 30 per cent over the next few years. In Europe seeds account for 20 per cent of the cost of producing wheat, compared with 45 per cent for fertilizers and 35 per cent for sprays. The share of seeds is expected to rise to 40-50 per cent by the year 2000.[20]

An additional effect of the new seed varieties is likely to be reorganization of labour relations in the Third World countries. Already, some breeding programmes have been focusing on producing crops that can be harvested by machines, which may make sense in the industrialized countries where labour costs are high. But if such plants are introduced in the Third World countries, they are likely to have major effects on the agricultural labour force. The shift in production relations is likely to affect women more severely because they constitute most of the agricultural labour force in these countries. Some of the new seed varieties are also more likely to be suited to large-scale production and will therefore displace smallholders or lead to the loss of their land to large farmers as happened during the Green Revolution.

Breeding programmes in countries such as the US are starting to shift away from minor crops and direct efforts towards the already established crops. This suggests that work on crops such as millet and sorghum will be undertaken only if there is a perceived market. 'The problem of minor crops is very similar to that of orphan drugs: The market is either not large enough to be profitable or not profitable enough. As a result, the private sector invests most heavily in major crops and, through grants, contracts, consulting, and even legislative means, pressures the public sector to focus on these crops.'[21]

This shift is part of a long transition in the growth of the breeding industry and seed companies. The seed business had been one of the most complex economic activities in the post-War period. While mechanical inventions and the resulting industrial operations have had clear positions in the market and the required intellectual property protection guaranteed, seed companies have had to deal with a much more uncertain situation. In the first place, the basis of the industry is a material that can be easily available and therefore trade secrets are not reliable as a measure of protection. Under such conditions, monopoly profits are difficult to guarantee and this has forced the industry to use innovation as one of the main sources of intellectual property protection.

The developments are part of a wider process of industrial reorganization in the field of biotechnology that are occurring at a time when the seed industry is consolidating itself. This trend is significant because the raw material for biotechnology is genetic resources. The significance of seeds in corporate strategies is illustrated by the diversity of firms participating in the industry and the intense competition to accumulate technological capability in genetic engineering, tissue culture and cell fusion. Ultimately, control over this resource will determine the level of entry into the market and access to the available resources. Gene banks have become a subject of controversy in recent years because they represent the most obvious control over genetic material.

Gene banks and food security

One of the most significant developments in world agriculture has been the establishment of a global network of gene banks. Modern gene banks store material under three types of germplasm collections. The first category is that of working collections, which are grown out annually by breeders and are usually kept at an ambient temperature or in airtight rooms. The second category includes medium-term collections which are dried and kept at sub-zero temperature (usually 0°C to -5°C); under such conditions, accessions are expected to last up to 30 years. The final category is that of long-term storage in which accessions are dried, sealed in airtight containers and stored at −20°C. Long-term accessions are rarely distributed or exchanged and are expected to last for more than 100 years. Germination tests are conducted on the accessions to ensure their viability. In addition to these categories, root crops are preserved in field plots or in tissue culture.

The evolution of the global gene banks is closely linked to plant collection activities either for taxonomic studies or for breeding. Since genetic resources are unevenly distributed and are moved from one place to another, it was almost inevitable that some form of storage would evolve. The mode of such storage systems, however, was largely influenced by the prevailing social and political conditions as well as the technological facilities available at the time. It is notable that the concept of gene banks is also consistent with the abstractionist principles that have governed Western thought since the 17th century. One of the realizations of such an approach is the isolation of key inheritable characteristics in genetic resources.

A further level of abstraction could be achieved through the isolation of the carriers of these genetic characteristics from the natural environment and locating them in a gene bank where ease of access and control could be ensured.[22] It should be noted that gene banks need not necessarily emerge because of such imperatives. Indeed, long-term germplasm storage was first undertaken in the Soviet Union in the 1920s at the All-Union Institute of Plant Industry in Leningrad under the leadership of Nikolai Vavilov. The accessions, however, were used largely for scientific and research purposes and the collection efforts were not closely linked to agricultural production. It was through Vavilov's large collection that he identified the major world centres of genetic diversity. As noted elsewhere, this work was later hampered by the rise of 'Social Lamarckianism' under the guidance of Lysenko.[23]

While Stalin was busy purging those geneticists who shared Vavilov's views in the 1940s, the US was entering a crucial phase in the evolution of its germplasm collection and plant introduction activities that started more than two centuries earlier. Four centres were established at Ames, Iowa (1974), Geneva, New York (1948), Experiment, Georgia (1949) and Pullman, Washington (1949) to introduce, multiply, evaluate, distribute and preserve germplasm either as seed or living collection. In addition, the Inter-Regional Plant Introduction Station for Potatoes was established at Sturgeon Bay, Wisconsin in 1949. These centres were equipped with cold storage facilities after it was found out that 90 to 95 per cent of the accessions recorded since 1898 when the inventory of introduced plants was established were no longer available in living collections.[24] In 1959, the US Department of Agriculture established the National Seed Storage Laboratory (NSSL) at Fort Collins, Colorado. This was the first US national gene bank.

Similar concerns led to the formation of a global network of gene banks at the international agricultural research centres (IARCs) of the Consultative Group on International Agricultural Research (CGIAR).[25] Breeders and scientists replenished their germplasm by going back to the source and collecting more material. But in the 1950s, it became obvious that genetic resources were being eroded and there was a need to conserve the collections in gene banks. The rate of genetic erosion in the Mediterranean region and the Near East, a centre of diversity for major cereals such as wheat, was alarmingly high, and the need for the conservation of the already collected material became an urgent issue. These concerns coincided with increasing breeding activities and the search for HRVs was already a major preoccupation of research institutes.

Table 3.2
CGIAR-supported research centres

Centre	Location	Programmes	Focus Budget	(US$ million)
IRRI (1960)	Los Baños	Rice	Global	22.5
	Philippines	Rice-based cropping systems	Asia	
CIMMYT (1966)	Mexico City,	Maize	Global	21.0
	Mexico	Bread wheat	Global	
		Durum wheat	Global	
		Barley	Global	
		Triticale	Global	
IITA (1967)	Ibadan,	Farming systems	Tropical Africa	21.2
	Nigeria	Maize	Tropical Africa	
		Rice	Tropical Africa	
		Sweet potato, yams	Global	
		Cassava, cowpea, lima bean, soybean	Tropical Africa	
CIAT (1968)	Cali,	Cassava	Global	23.1
	Colombia	Field beans	Global	
		Rice, tropical pastures	Latin America	
CIP (1971)	Lima, Peru	Potato	Global	10.9
WARDA (1971)	Monrovia, Liberia	Rice	West Africa	2.9
ICRISAT (1972)	Hyderabad, India	Chickpea, pigeon pea, pearl millet, sorghum, groundnut	Global	22.1
		Farming systems	Semi-arid tropics	

Table 3.2 cont.
CGIAR-supported research centres

Centre	Location	Programmes	Focus Budget	(US$ million)
ILRAD (1973)	Nairobi, Kenya	Trypanosomiasis Theileriosis	Global Global	9.7
IBPGR (1974)	Rome, Italy	Plant genetic resources	Global	3.7
ILCA (1974)	Addis Ababa, Ethiopia	Livestock production systems	Tropical Africa	12.7
IFPRI (1975)	Washington, DC, USA	Food Policy	Global	4.2
ICARDA (1976)	Aleppo, Syria	Farming systems Wheat, barley, triticale, lentil, chickpea, broad bean, forage crops	Dry areas of West Asia and North Africa	20.4
ISNAR (1980)	The Hague, Netherlands	National agricultural research	Global	3.5

Source: Consultative Group on International Agricultural Research, Washington, DC

The origins of the global network of gene banks is closely associated with the internationalization of agricultural research. The early results of research on HRVs in Mexico and Asia illustrated the importance of making available a wide range of genetic material on which to base breeding programmes. By the late 1950s FAO was playing a major role in the promotion of plant breeding activities, especially in the dissemination of information. Like many UN agencies of the day, FAO worked closely with the private sector in facilitating research and promoting new varieties and technologies in the Third World countries. By then, FAO was already active in the dissemination of information on genetic resources. Its *Plant Introduction Newsletter* was designed to make available information on the collected germplasm and promote seed exchange as well as plant introduction.[26]

Following a 1961 FAO technical meeting, it was felt that a panel of experts should be formed to work on plant exploration and introduction. Thus, activities which hitherto had been undertaken by individuals and governments were now to be conducted under the auspices of an international body. Indeed, the requirements for such activities made it necessary to operate through an international institution. A panel of experts was set up in 1965 and further conferences on the subject were organized in 1968 and 1973. The 1973 conference made specific proposals for the exploration, collection, conservation, documentation, and evaluation of plant genetic resources. The meeting also proposed that an international network of gene banks be established. A year earlier, the

Conference on the Human Environment in Stockholm had called for a global programme to conserve tropical and sub-tropical crop genetic resources.

This period marked a significant change in the institutional organization of plant genetic resource activities. While FAO undertook most of the initial work, a new autonomous institution, the International Board for Plant Genetic Resources (IBPGR) was set up in 1974 to deal with plant genetic resources; views differ on how this change occurred. One view holds that the transition occurred because the emergent CGIAR, an off-shoot of the Green Revolution, usurped the activities of the Plant Ecology Unit of FAO.[27] The other view holds that the transition occurred because of the inability of FAO to facilitate the exploration, collection, conservation, documentation and evaluation of plant genetic resources. 'The fact is that IBPGR would not have been conceived had FAO shown strong determination for concerted action to establish a genetic resources program. IBPGR originated not in competition, but out of frustration with FAO.'[28]

In order to understand the factors that led to the shift of control from FAO to IBPGR, it is necessary to examine the institutional reforms that were occurring at the time and the related changes in the world agriculture. The Green Revolution was starting to show its success in the rising gross output of grain, and international donors such as the World Bank, Rockefeller Foundation and Ford Foundation were consolidating and co-ordinating their work. This collaboration led to the formation of the CGIAR following the outbreak of the corn leaf blight in the US which enhanced awareness on the need for germplasm conservation. CGIAR's underlying organization philosophy was networking. It is, therefore, not surprising that scientists and donors argued for a network of gene banks and not centralized facilities. For the CGIAR to guarantee long-term operations, the need for access to germplasm was crucial. Leaving this work in the hands of FAO would have been risky, especially in view of the fact that FAO was an international agency accountable to governments and susceptible to red tape. A new institution was thus needed.

It was, however, necessary for such an institution to build on the experience and expertise accumulated by FAO over the years. One way of achieving this was to create an institution that would operate in close association with FAO, sharing its institutional niche and all its benefits as well as privileges of access, while at the same time ensuring that it serves the interests of the sponsoring organizations, through the CGIAR. This, indeed, was a very effective way of operating and most of the objectives of IBPGR for the first decade have been achieved.[29]

In the first place, IBPGR had full access to the knowledge and expertise built by FAO over the post-War period. Second, IBPGR was able to operate as if it was a UN agency having full access to its member countries. Third, under this arrangement, IBPGR was not directly accountable to the members of FAO. This uncertain status of the institution raised questions on its legitimacy as well as the ownership of the genetic resources collected under its auspices. The question of ownership relates not only to access by countries of origin, but also deals with the legal status of agreements entered between IBPGR with the conserving banks, especially national facilities which are subject to national law.[30]

Table 3.3
Estimate of global gene bank accessions

	Number of			Coverage Percentages	
Crop	**Accessions**	**Distinct Samples**	**Collections of More than 200 Accessions**	**Landraces**	**Wild Species**
Cereals					
Wheat	410,000	125,000	37	95	60
Barley	280,000	55,000	51	85	20
Rice	215,000	90,000	29	75	10
Maize	100,000	50,000	34	95	15
Sorghum	95,000	30,000	28	80	10
Oats	37,000	15,000	22	90	50
Pearl Millet	31,000	15,500	10	80	10
Finger millet	9,000	3,000	8	60	10
Other millets	16,500	5,000	8	45	2
Rye	18,000	8,000	17	80	30
Pulses					
Phaseolus	105,500	40,000	22	50	10
Soybean	100,000	18,000	28	60	30
Groundnut	34,000	11,000	7	70	50
Chickpea	25,000	13,500	15	80	10
Pigeon pea	22,000	11,000	10	85	10
Pea	20,500	6,500	11	70	10
Cowpea	20,000	12,000	12	75	1
Mung bean	16,000	7,500	10	60	5
Lentil	13,500	5,500	11	70	10
Faba bean	10,000	5,000	10	74	15
Lupin	3,500	2,000	8	50	5
Root Crops					
Potato	42,000	30,000	28	95	40
Cassava	14,000	6,000	14	35	5
Yams	10,000	5,000	12	40	5
Sweet potato	8,000	5,000	27	50	1

Table 3.3 cont.
Estimate of global gene bank accessions

Crop	Accessions	Number of Distinct Samples	Collections of More than 200 Accessions	Coverage Percentages Landraces	Wild Species
Vegetables					
Tomato	32,000	10,000	28	90	70
Cucurbits	30,000	15,000	23	50	30
Cruciferae	30,000	15,000	32	60	25
Capsicum	23,000	10,000	20	80	40
Allium	10,500	5,000	14	70	20
Amaranths	5,000	3,000	8	95	10
Okra	3,600	2,000	4	60	10
Eggplant	3,500	2,000	10	50	30
Industrial Crops					
Cotton	30,000	8,000	12	75	20
Sugar-cane	23,000	8,000	12	75	20
Cacao	5,000	1,500	12	*	*
Beet	5,000	3,000	8	50	10
Forage					
Legumes	130,000	n.a.	47	n.a.	n.a.
Grasses	85,000	n.a.	44	n.a.	n.a.

* Coverage difficult to estimate because many selections are from the wild.
n.a. = data not available.
Source: Plucknett et al., *Gene Banks*, pp. 111-112.

From the available statistics, the global gene bank network represents a major effort in relation to the budgetary limitations of IBPGR. It is estimated that global expenditure on crop genetic resources by the mid-1980s was US$55 million of which some US$3.8 million (6.9 per cent) went to IBPGR. The bulk of the funding, 53 per cent, was spent on national programmes in the industrialized countries (mainly the US) while the Third World countries spend only 14 per cent of the total budget. The IARCs accounted for 17 per cent and the rest was spent by bilateral aid agencies, foundations and multilateral bodies (mainly UN agencies).[31]

Despite these limitations, the contribution of gene banks to the global agriculture is considerable. The benefits have been in the form of providing genetic material with disease-resistant, pest-resistant, and environment-tolerant characteristics. The varieties that dominate global food production such as the IR-36 rice developed by IRRI benefited partly from material that was previously stored in gene banks.[32] The contributions have, however, tended to be largely based on vertical resistance which is race-specific and relies largely on single-gene inheritance, a truly Cartesian research programme. This means that if the pathogen undergoes distinctive mutation, it can lead to new races which can affect the crop. The effects of an attack could be severe because of the predominant monocultural practices associated with the Green Revolution. This is the soft underbelly of uniformity. After widespread attacks by pests and disease, IRRI is now shifting some of its programmes to horizontal resistance, an approach that requires access to an even larger pool of genetic material.

The history of the gene bank network has its political dimensions, some of which border on institutional intrigue. While most of the institutions have relied heavily on the collection of material from the wild and the exchange of germplasm from other institutions, some have used political pressure to take over the germplasm collected over the years by other institutions as well as the associated personnel. This is exemplified by the conflict between IRRI and the Central Rice Research Institute (CRRI) at Cuttack, India, in the early days of the Indian rice Green Revolution. The institutional base of IRRI and its influence over politicians in India made it difficult for national research programmes to survive its assault. The changing of the Indian rice research programme involved the retirement of those who refused to follow the IRRI-type of research; those who opposed the reductionist methodology. A leading casualty in this process was R. H. Richharia who was unceremoniously retired from the CRRI.[33] His opponents claimed that he was not a scientist. In so far as he rejected the reductionist methodology, Cartesian scientists did not consider him one of their kind. He had to be derided.[34] Baconian imperialism was on the march and lone scientists could not stop its expansionism.

The supplanting of major national research programmes by IRRI's research agenda and varieties was not the end of the story. In the 1970s IRRI increased its interest in collecting rice germplasm, mainly to expand the genetic base of their breeding programme in response to the low pest-resistance in the new varieties.[35] Richharia, who had been retired from CRRI, partly because of IRRI's pressure, was now in control of over 19,000 varieties of rice conserved *in situ* at the Madhya Pradesh Rice Research Institute (MPRRI) at Raypur.[36] Not only was he in control of traditional varieties that yielded 8-9 tonnes per hectare, higher than some of IRRI's improved varieties, he had also discovered dwarf plants without the susceptible dwarfing gene that IRRI had incorporated in its varieties.[37] When MPRRI refused to share the material with IRRI on the grounds that the plants had not yet been adequately studied, IRRI reportedly engineered the shutting down of the institute. The material from the institute was later handed over to the Jawaharlal Nehru Krishi Vishwa Vindyalaya (JNKVV) centre from where it was transferred to IRRI.

It should be noted that MPRRI's work was not in line with the long-term objectives of corporate breeding programmes. MPRRI had already started to release disease-resistant and pest-resistant varieties – a line of breeding likely to undermine the sale of pesticides which relied heavily on the susceptibility to disease of the improved rice varieties, like those released by IRRI. It was partly for this reason that MPRRI was reluctant to exchange its material with IRRI: they had a different research agenda and even feared that the disease susceptibility in IRRI varieties might be introduced in Madhya Pradesh. While at IRRI, Richharia had questioned IRRI officials bringing rice into the country without quarantine certification, fearing that such practices could lead to the introduction of rice diseases into the country. The fight against Richharia was vicious: not only did the establishment take away his breeding material, but on the two occasions they drove him from office, they also confiscated his research papers and notes.

This struggle should be placed in the context of historical developments in India at the time. The country experienced low food output over the 1965-66 period due to the failure of the monsoon rains to arrive in time. The ensuing food shortages forced India to import some 10 million tonnes of wheat from the US. With the spectre of food shortages hanging over the Indian nation, government officials were more willing to accept the introduction of wheat and rice HRVs. It so happened that the shortages occurred at the time when the international research system had new varieties available and the prevailing conditions made their widespread use more feasible. The pressure was immense, and scientists like Richharia who stood in the way had to be removed to allow those who favoured the designs of the promoters of the Green Revolution to take over.[38] Richharia's work was associated with the low-response varieties that improved varieties from IRRI were poised to replace; it was difficult to replace such varieties if some leading scientists were still producing them. With the removal of Richharia, the Cartesian programme carried the day.

Not only did the enormous power of the international research institutions become an issue of concern but questions pertaining to access to the material they held became a matter of controversy. It is notable that some of these are base collections which are not routinely distributed. 'Materials are only moved from base collections in order to regenerate materials when the seed viability has started to decline below an acceptable regeneration standard, or when stocks of a particular sample are no longer available from other sources. *Seed is not normally used for routine distribution from a base collection.*'[39] Some of the regenerated seeds are placed in an active collection where they can be obtained for distribution, evaluation and multiplication.

There are numerous ways in which seeds are distributed to the Third

World countries. One of the channels is through US Agency for International Development (AID) projects. Other channels include the US Department of Agriculture's Agricultural Research Service (ARS) Plant Introduction Office at Beltsville, Maryland. The office has maintained a tradition of exchanging germplasm with various countries and research institutes. Being a major centre in the collection and introduction of germplasm, the office has maintained good relations with the Third World countries.

Table 3.4
Seed accession from the USDA-ARS Beltsville office

Country	1975	1981	1982	1983	1984	1985
Angola	—	—	—	1	11	—
Botswana	11	2	—	89	78	15
Burkina Faso	4	—	35	73	325	36
Burundi	15	—	—	—	—	5
Cameroon	82	3	34	33	45	9
Canary Islands	—	—	—	—	—	12
Cape Verde	—	—	—	47	—	—
Chad	—	—	—	1	—	—
Congo	1	—	—	—	—	—
Egypt	944	778	1,229	392	194	887
Ethiopia	989	912	418	831	553	419
Gabon	—	—	11	2	—	—
Ghana	48	17	5	—	116	90
Ivory Coast	11	351	—	18	15	10
Kenya	5,040	2,790	1,341	1,149	2,035	246
Lesotho	50	—	25	—	—	—
Liberia	3	—	—	12	—	—
Libya	—	—	—	6	—	—
Madagascar	15	5	—	—	3	—
Malawi	28	9	19	3	2	15
Mali	262	3	48	—	3	—
Mauritania	—	9	3	18	3	—
Mauritius	—	—	2	4	27	109
Morocco	12	240	411	441	58	194
Mozambique	11	1	5	5	—	10
Namibia	—	5	—	—	—	—
Niger	—	106	31	4	—	6
Nigeria	3,928	376	362	4	144	149
Rwanda	—	—	9	4	—	25
Senegal	23	1	18	3	5	39
Sierra Leone	—	29	43	17	48	6
Somalia	—	—	28	1	3	2
South Africa	2,423	3,437	6,657	4,661	4,147	3,743
Sudan	70	178	113	34	63	13

Table 3.4 cont.
Seed from the USDA-ARS Beltsville office

Country	1975	1981	1982	1983	1984	1985
Swaziland	18	—	—	—	—	6
Tanzania	669	29	24	33	160	51
Togo	—	2	1	—	59	4
Tunisia	80	700	346	353	108	75
Uganda	130	519	2	1	—	1
Zaïre	10	25	55	30	8	4
Zambia	669	25	574	228	89	2,974
Zimbabwe	—	995	511	862	973	281

Source: USDA-ARS, Plant Introduction Office, Beltsville, Maryland.

The other area of legal concern relates to genetic resources in private collections. It can be assumed that such material is private property and governed by the legal instruments that protect such property. In addition to this, it is notable that information pertaining to genetic resources in private collections is not as easily available as that on material in public sector gene banks. Genetic resources collected by seed companies are not likely to be freely exchanged. Moreover, these resources may be assembled for specific breeding programmes and information pertaining to them is not freely made available as it may constitute trade secrets.

So far there are only a few cases where Third World countries have reportedly been denied access to germplasm collected from their territories. Moreover no Third World country has legally attempted to repatriate any genetic resources from a foreign country. Seed repatriation has been suggested as a way to legally enable the Third World countries to have access to resources originally collected from their territories. The concept is based on debates at UNESCO in the 1960s which led to the signing of the UNESCO Convention on the Means of Prohibiting and Preventing Illicit Import, Export and Transfer of Ownership of Cultural Property in 1970.

Following the convention, the UK returned Kenya's two-million-year-old *Procunsul africanus* skull. Other countries such as France, the US, South Africa, Belgium and the Netherlands have repatriated cultural artefacts to their countries of origin. There have been attempts to treat seeds as cultural property and therefore justify their repatriation.[40] Ironically, a recent case of seed repatriation occurred in Ethiopia in 1986 as an initiative of the ARS. In 1967 ARS, the East African Agricultural and Forestry Research Organization (EAFRO) and the University of Nebraska (US) collected 138 samples of teff (*Eragrostis tef*), a native plant of Ethiopia used to make a pancake-like bread which is a staple in the country.

The resources were in cold storage in the US until 1985 when they were replanted for revitalization. The ARS sent an inventory of the teff collection to the Debra Zeit Research station in Ethiopia which had earlier added 170 samples to the American collection. On receipt of the inventory, the Plant Genetic Resources Centre (PGR/E) in Addis Ababa requested seed, which was subsequently delivered to Ethiopia by a senior ARS official.[41] Teff has no significant commercial value in the US although other countries have shown interest in developing it as a commercial crop; Wye College of the University of London and the Volcani Centre (Israel) have been working on the plant, and FAO has implemented a research programme at the Holleta Research Station near Addis Ababa.[42] Attempts are underway to develop dwarf varieties less susceptible to disease and pests. The plant tends to bloom for a short time after sunrise, making it difficult to cross-pollinate it before the flower closes.

The decision by ARS to return teff seeds to Ethiopia shows the significance of access to information on collected resources and willingness of the depositories to return the material to the country of origin. It should be noted that no legal issues were invoked in the repatriation. It was in the interest of the US to guarantee that international germplasm exchange is not restricted, especially with countries such as Ethiopia which are centres of genetic diversity. At the time of the delivery of the teff seeds, the ARS was interested in Ethiopia's collection of *Vernonia galamensis* varieties, a plant currently being tried in Zimbabwe as a potential source of epoxy oils for use in the manufacture of plastics and protective coatings. The plants growing in Zimbabwe are based on genetic resource originally obtained from Ethiopia, which still holds accessions collected in 1964 and necessary for *Vernonia* breeding programmes. In the 1970s, a Kenya-based firm, African Highlands Produce, grew *Vernonia* using seed from Ethiopia, and exported seeds to the US as a source of coatings.[43] The economic interest in *Vernonia* required increased access to germplasm.

The development of *Vernonia* as an industrial crop highlights most of the controversial issues related to the international flow of germplasm. Although the African countries are the main sources of *Vernonia* germplasm, the commercial production of the crop may take root in the US. Although seed for research is now being produced in Zimbabwe, future sources may be located in the US. The current advances in biotechnology also offer the possibility of inserting the genes that code for the production of vernolic acid into established crops whose genetic structure is well known. This will help to reduce the long period required for agronomic studies and breeding. Except for the Zimbabwean trial farms, there are no African programmes to look into the potential for *Vernonia* as an industrial crop.

Concern over the legal status of *ex situ* material has been discussed in detail at the FAO Commission on Plant Genetic Resources and the likely course of action seems to be to bring the material collected by IBPGR under the control of FAO as a custodian. The measures to achieve this objective have already been initiated by the setting up of the International Fund for the Conservation and Utilization of Plant Genetic Resources. Putting the resources under the control of FAO would enable the Third World countries to legally have access to the material. The genetic resources, however, still remain a source of legal conflict for some time to come.

There are other conflicts in this field. The source of funding for gene banks in the Third World countries is becoming a major problem in relation to access. In addition to the IARC gene banks, some facilities in the Third World are funded through bilateral aid programmes. It is expected that the main donor country has direct access to the material held therein. In some cases the gene banks are funded with the understanding that the collected germplasm will be shared with the donor country. Since most such arrangements are made under the banner of scientific and technical co-operation, issues of sovereignty do not normally feature in the initial stages of negotiation. Only after the gene banks start operating and the economic value of the material starts to be salient does the question of sovereignty start to emerge.

This issue is exemplified by the conflict between Ethiopia and West Germany over barley genetic resources. The PGR/E in Addis Ababa was built with the help of funds from the Deutsche Gesellschaft für Technische Zusammenarbeit (GTZ), a West German technical co-operation agency. Over the years GTZ has continued to provide Ethiopia with the required financial support to run the bank but its management and control have remained firmly in the hands of Ethiopians. In return for the support, Germany had required Ethiopia to provide it with duplicates of its collections, especially of barley. An official sent to Ethiopia by GTZ in 1985 to negotiate for the delivery of the duplicates returned home in 1987 empty-handed. The GTZ reportedly responded by refusing further support to the gene bank.

Ethiopia's decision to refuse the delivery of duplicates was largely based on safeguarding seed sovereignty. Government officials, however, have argued that the accessions are 'populations' and therefore no duplicates can be made. They stress that sharing the seeds will leave each party with different samples and not duplicates. The Germans, on the other hand, have argued that it is risky to keep samples in one place without duplicates. This is ironic because the Institut für Pflanzenbau und Pflanzenzüchtung der Bundesforschunsanstalt für Landwirtschaft (FAL) gene bank at Braunschweig has failed to provide adequate duplicates of their designated IBPGR collections to other gene banks.

They have refused to do so despite the fact that they have committed themselves to providing duplicates. Aid programmes have been a significant medium through which germplasm has been collected from the Third World countries. The case of Ethiopia illustrates the delicate problems that might arise if Third World countries attempt to assert their sovereignty over their botanical heritage.

One of the main concerns about the ability of gene banks to maintain food security over the long run has been the security of the gene banks themselves. Ever since the debate on gene banks started in the late 1970s, there has been growing concern over the ability of these institutions to guarantee the safety of their accessions. Data from various FAO, IBPGR and government reports in the early 1980s revealed sporadic losses of germplasm in gene banks.[44] The reports give a wide range of causes for the losses, including negligence, poor storage conditions, equipment breakdown, fire, disease, and loss of viability. By the early 1980s the US government admitted that the national gene bank network, with over 400,000 accessions, had inadequate storage facilities.[45] The situation in the US has become worse, especially as a result of the Reagan budget cuts.

These concerns have put in focus NSSL, the largest gene bank in the world, which has over 215,000 accessions and was running out of space by 1987. With a professional staff of nearly a dozen people, the bank could not cope with the work, especially growing out the material and testing for viability. If the world botanic equivalent of Fort Knox was having trouble, it was easy to imagine what could happen in other gene banks elsewhere. It was estimated that by 1986 nearly 100,000 accessions in the US alone badly needed to be grown out but financial and personnel limitations could not allow the exercise to be undertaken. The US case alone, however, could not be used as an indicator of the state of the global gene bank network. It could also be argued that the data collected in the early 1980s was anecdotal and, therefore, could not be used to pass judgement on the network.

The state of safety in international gene banks has not improved since the concerns of the early 1980s. A study presented to the IBPGR Board of Trustees in 1987 revealed major inadequacies in the 'designated base' gene banks.[46] According to the study, seven of the 17 evaluated banks did not meet IBPGR's registration standards. In addition to these, several other gene banks requested that the evaluation be delayed or did not respond to the letters requesting them to be evaluated. This suggests that their standards, too, may have been unacceptable. Although the number of banks so far evaluated is small, they represent 50 per cent of available storage space (in the survey) and 60 per cent of surveyed germplasm.[47] The seven banks account for 13 per cent of the world's stored germplasm.[48]

A preliminary ranking of the performance of the gene banks suggests that Third World banks are more effective than industrialized country counterparts. This performance lends support to the need to strengthen the capacity of the Third World countries to conserve their resources. Moreover, the costs of gene bank management, especially those relating to personnel, may be lower in the Third World than in the industrialized countries. The proximity of these gene banks to the centres of diversity gives them comparative advantage over their northern counterparts. What is needed, though, is an effective training programme to equip the Third World countries with the technical ability to manage even more effectively gene banks at regional, national and local levels.

Table 3.5
Ranking IBPGR evaluation of 17 gene banks*

Country	Institution	Status	Rank
CGIAR/IARC	IRRI, Los Baños	Acceptable	1
United Kingdom	NURS, Wellesbourne	Acceptable	2
Ethiopia	PGRC, Addis Ababa	Acceptable	3
United Kingdom	RBG, Kew	Acceptable	4
Nigeria	IITA, Ibadan	Acceptable	5
Taiwan	AVRDC, Taiwan	Acceptable	6
Italy	CNR, Bari	Acceptable	7
Thailand	TISTR, Bangkok	Acceptable	8
Australia	CSIRO, Canberra	Unacceptable	9
Australia	CSIRO, Samford	Unacceptable	10
Federal Republic of Germany	FAL, Braunschweig	Acceptable	11
CGIAR/IARC	ICARDA, Syria	Unacceptable	12
USA	NSSL, Fort Collins	Unacceptable	13
Greece	GGB, Thessalonika	Unacceptable	14
Spain	Polytech, Madrid	Unacceptable	15
Nordic	NGR, Lund	Acceptable	16
Canada	PGRC, Ottawa	Unacceptable	17

*This is a RAFI ranking of the quality of gene banks.
Source: RAFI, '*Security of Gene Banks*,' p. 9.

In this evaluation IBPGR has confirmed that the gene banks, in their current state, cannot be relied upon to guarantee food security over the long run. Much needs to be done to ensure that the banks are adequately funded and are less affected by technical and administrative problems. The problems of some of these banks go beyond the accessions; some simply do not keep their files in order. At the time of the evaluation the curator of the Plant Gene Resources Centre of Canada office at Ottawa, for example, did not have copies of the original agreement between Canadian authorities and IBPGR and had to request for copies from Rome.

On the whole, the success of recent efforts to set up international institutions that co-ordinate agricultural research and germplasm conservation has led to numerous controversies. Part of the problem is that the research activities are based on philosophical guidelines and practical experiences arising from certain historical circumstances and may be unsuited to other parts of the world. The research programme has been largely antithetical to diversity and has therefore run into serious problems when confronted with such conditions. Moreover, the social content of the research agenda does not necessarily conform to the aspirations of a large section of the world population. Agriculture has, over the centuries, been a process through which people (individually or collectively) are linked to the natural environment in their daily lives in a sustainable way. With the advent of industrialization, this process, especially with the application of the mass production paradigm, has lost that basic characteristic. The social impact of modern agriculture has also been closely associated with major, irreversible changes in the ecology.

The changing economy of nature

The introduction of any new crops changes the existing crop ecology. The magnitude of the change depends largely on the nature and extent of the new plants as well as the natural resource base. One of the features of the introduction of HRVs was the reduction in the use of traditional varieties. As a result of the introduction of HRVs some varieties that were retired from use were lost. In the early days of the Green Revolution this loss did not seem to concern the international community as it did later. In some cases HRVs were introduced in conjunction with the planned reduction in the use of indigenous varieties. This was the Japanese strategy in Taiwan and was practised by other colonial powers.

The loss of indigenous genetic resources due to the introduction of HRVs, especially if used in monocultures, has been a major source of criticism against the Green Revolution; there have been two main responses to this criticism. The first has been to ask for more evidence, the typical Baconian demand for objectivity; the other has been to collect the available resources and store them in gene banks. The defence of the Green Revolution, however, has been able to abate fears that the world is becoming poorer in genetic diversity as uniform varieties are introduced.

In some cases, the Green Revolution undermined its own capacity to respond to some of the effects of its successes. The case of rice breeding in the Philippines illustrates the point. When the IR-8 variety was attacked by the tungro disease, the farms switched to IR-20, but this hybrid proved vulnerable to grassy stunt virus and brown hopper insects.

The farmers were then supplied with the IR-26 hybrid which appeared to be resistant to most of the diseases and pests in the country but it proved vulnerable to strong winds. When breeders decided to try the original Taiwanese variety that withstood strong winds, they found that it had been lost as Taiwanese farmers planted their farms with IR-8.[49]

Genetic vulnerability has been a major challenge for plant breeders. The problem has been manifested by the rise of previously insignificant pests or diseases into major crop epidemics as a result of the genetic vulnerability of major crops. Over the 1969-70 period, a virulent strain of southern corn blight raised a major scare in the US and helped stimulate some of the recent plant collection activities. The previously undetected race of *Helminthosporium maydis* is estimated to have destroyed 15 per cent of the US corn, an estimated loss of US$15 billion.[50] By then, nearly 96 per cent of the peas produced in the US were from only two varieties. Over that period, US plant hunters combed the world for peas with resistance to cold, drought and disease. It is through such activities that some of the major US crops have been saved from collapse. By the late 1960s, the vulnerability of the US and Canadian agriculture was reflected in the narrowness of their agricultural base: only four varieties accounted for 75.9 per cent of the bread wheat grown in Canada; nearly 69 per cent of all the sweet potatoes grown in the US were based on one variety; all the millet in the US was from three varieties; and 71 per cent of the corn was dependent on six varieties.[51]

In the early 1980s, US farmers had, indeed, improved the genetic diversity of their crops. This was partly as a result of the 1970 scare and the concerted efforts that went into plant collection and the establishment of gene banks. The diversification of the genetic base in the US went a long way in introducing a measure of security in agricultural production. There was a marked decline in the concentration of major cultivars, except for wheat, as farmers expanded the genetic base for their production.[52] In addition to the diversification, the decade had also experienced major shift from one variety to another – genetic diversity in time. It is doubtful whether genetic diversity in time represents an improvement in crop breeding or is a result of the pressure on plant breeders to introduce new varieties through minor modifications.

It should be noted that genetic vulnerability is related not only to uniformity but also to methods used in plant breeding. The southern corn leaf blight resulted from the hybridization methods used hitherto. The corn then was largely F_1 hybrids consisting of female patent lines carrying Texas male sterile cytoplasm. The lines are male-sterile because of the changes introduced in the mitochondrial DNA in comparison with fertile lines. This method ensured that only F_1 hybrid seed was produced on the rows of female parent plants in seed production farms. Since they do not produce pollen themselves, they are pollinated only by the nearby sterile male parent. This was an effective way of producing hybrids but it also made corn susceptible to *H. Maydis*. The use of cytoplasmic male sterility (CMS) was abandoned by some seed producers who now use mechanical detassling to remove terminal male flowers before they release pollen.[53]

Interestingly enough, scientists were aware of the potential effects of *H. maydis* on corn based on inbred lines from the US containing Texas male-sterility cytoplasm ('T-cytoplasm') in the 1960s but were not able to take any preventive measures. Philippino scientists published studies in 1962 and 1965 showing that *H. maydis* had destroyed their hybrid crop as early as 1957. Scientists at Pioneer Hi-Bred and the University of Illinois checked the Philippino studies in 1965 but came to the conclusion that the effects of *H. maydis* must have resulted from secondary effects of reduced plant vigour or factors specific to the local environment. This case illustrates two important points: first, private corporations which have invested large sums of money are not usually interested in verifying claims on problems that might require major changes in their productive techniques. They tend to do so only when faced by overwhelming evidence as it happened in 1971; second, the Cartesian worldview may lead to the conclusion that different countries constitute separate environments and cannot, therefore, be used to draw conclusions on possible impact in other countries.[54]

The agricultural model of the last several decades has also been associated with the intensive use of pesticides and fertilizers whose human and environmental effects have been widespread.[55] As will be shown in the next chapter, the issue of ecological change has entered a new phase of uncertainty, especially in relation to the major advances that have occurred in the field of biotechnology. Some of the innovations are aimed at reducing the release of agrochemicals into the environment. The release of genetically engineered micro-organisms into the environment is, however, posing new challenges for which most of the institutions are not prepared. This is just one of the discontinuities being introduced by advances in biotechnology.

Another significant problem associated with the Green Revolution is the drain of micro-nutrients from the soil – a problem that is not well documented and has, therefore, received little attention in the literature. Farmers tend to add potassium, nitrogen and phosphorus back into the soil although HRVs use up other micro-nutrients that are essential for the healthy growth of the plants. These nutrients include zinc, copper, iron, manganese, magnesium, molybdenum and boron which, although they form about only one per cent of the plant weight, control numerous plant functions. Zinc, for example, helps plants to use nitrogen and phosphorus; its depletion, therefore, can undermine the capacity of crops to use these fertilizers. In various parts of India the depletion of micro-nutrients is starting to affect crop yields.[56]

This chapter has so far examined how the plant collection efforts of the 18th and 19th centuries led to the formation of a complex network of international agricultural research activities. The agricultural model that emerged from these developments was largely based on an agricultural system that resembled mass production. By the 1940s the US had developed a monocultural agricultural system that depended on a narrow range of genetic resources. This model was transferred to numerous Third World countries with major social, political and ecological implications. The process was also associated with the complex organization of the flow of genetic resources to meet the needs of agricultural research.

At the core of the Green Revolution was the expanded use and flow of genetic resources. This use was associated with the evolution of institutional networks and corporate entities which have a firm grip on key genetic resources used in the global agriculture. The world is entering a new phase in agricultural production in general and the use of genetic resources in particular. Biotechnology, as the collection of techniques is referred to, is likely to affect the Third World countries in ways that cannot be predicted.

Notes

1. The Green Revolution involved other major changes in the application of agrochemicals, and irrigation as well as land reform and social reorganization.
2. See Yamada, 1967, for additional details on this theme.
3. Ogura (ed), 1966, p. 320.
4. Pearse, 1980, p. 27.
5. Ibid., p. 28.
6. Piore and Sabel, 1984, p. 4.
7. IGRP, 1985, p. 29.
8. See Zirkle, 1949b, for a historical account of hybridization in the 18th century.
9. See Berland and Lewontin, 1986, for a review of such developments.
10. Belsky et al., 1985, p. 10.
11. For a more detailed discussion of UPOV see Chapter 5.
12. '[T]he U.S. seed industry is a distinctly different entity from that which prevailed in the late 1960s. It is more concentrated, more competitive, more technologically sophisticated, has a more favorable division of labor with the public research system, and is more internationally oriented. The industry is also more firmly anchored in intellectual property restrictions such as PVPA and patents at the same time that its emphasis has continued to revolve around reproductively unstable hybrid varieties,' Belsky et al., 1985, pp. 16-17.

13. The industry was built with the help of nearly 5,000 chemical patents seized by the US federal government from Germany during World War I, many of which belonged to the I.G. Farben complex. The patents were sold to the Chemical Foundation by the Alien Property Custodian, who later resigned from the government to become the President of the Foundation. Government legal action brought against the Foundation in the Supreme Court was defeated. Not only did the US chemicals industry exploit the patents, but the government also imposed a protective tariff against the products of a reviving German chemical trust. The 1922 tariff system remained in force until the 1964 Kennedy round of tariff negotiations. See Borkin, 1978; Juma, 1986; and Clark and Juma, 1987.

14. Bennett and Kline, 1987, p. 39.

15. Ibid., p. 44.

16. Monsanto's in-house research investment amounted to US$40 million in the mid-1980s. It had invested equity in biotechnology firms such as Collagen (artificial bone powder for biomedical research), Biogen (tissue plasminogen activator), Genentech (bovine growth hormone) and Genex (venture capital investment) and contracted Harvard University (US$23 million for biomedical research), Washington University (US$23.5 for biomedical research) and Rockefeller University (US$4.0 million for photosynthesis research).

17. Reacting to fears of ICI monopoly in the industry, a company spokesman was reported as saying: 'I don't think we will ever get to the stage where ICI can squeeze the balls of every farmer in the world, because every other chemicals company will be doing the same thing,' Erlichman, 1987.

18. The NSDO receives the rights to breed all the varieties that come from public sector institutions. 'The PBI creates most of NSDO's wealth. Over 80 per cent of the NSDO's revenue comes from royalties on varieties created at PBI. In the financial year 1984/85, the income of the NSDO was about £10.8 million. Of this revenue, the government received about £6 million, which went to the Treasury. None of this money went back to PBI, which is funded partly from the Science Vote of the Department of Education and Science and partly by commissions from the Ministry of Agriculture, Fisheries and Food. The full cost of running the PBI is less than £5 million. The profit that NSDO makes could therefore pay for the PBI', Connor, 1986, p. 35.

19. Rendezvous is a cross between wheat and a wild goat grass, *Aegilops vetricosa*.

20. *The Economist*, August 1986, p. 56.

21. 'Furthermore, with public funds available for biotechnology programs involving only a few crops, research is likely to follow the interests of the most powerful commodity associations, which tend to represent major crops. Research to increase the role of minor crops or to increase the number of food crops humans can use will likely be underfunded', Hansen, 1986, p. 36.

22. But such storage could be achieved only if the relevant technology was

available. 'The technology for long-term preservation of seeds was in place long before the emergence of modern gene banks. Machines for making ice and freezing meat were in use by the mid-1800s . . . In the 1920s, freon-based refrigeration equipment was developed, providing a more efficient and less dangerous way of storing goods; earlier systems relied primarily on ammonia, which could leak toxic fumes. Plant breeders and the seed trade were behind the push to store plant germplasm under long-term conditions; fortunately, technology was available when the need arose', Plucknett et al., 1987, p. 67.

23. See Popovsky, 1984.

24. Plucknett, 1987, p. 68.

25. The CGIAR was set up in 1971 and comprises 50 countries, international and regional bodies as well as private foundations. It is sponsored by the World Bank, Food and Agriculture Organization (FAO), and the UN Development Programme (UNDP). The secretariat and chairman of the CGIAR are provided by the World Bank. Its budget for core programmes for 1985 was US$170 million from 34 donors. The CGIAR supports 13 research institutes: International Rice Research Institute (IRRI), International Crops Research Institute for Semi-arid Tropics (ICRISAT), International Board for Plant Genetic Resources (IBPGR), Centro Internacional del Mejoramiento de Maiz y Trigo (CIMMYT), International Institute of Tropical Agriculture (IITA), Centro Internacional de Agricultura Tropical (CIAT), Centro Internacional de la Papa (CIP), International Centre for Agricultural Research in Dry Areas (ICARDA), International Livestock Centre for Africa (ILCA), International Laboratory for Research in Animal Diseases (ILRAD), International Food Policy Research Institutes (IFPRI), International Service for International Agricultural Research (ISNAR), and West African Rice Development Agency (WARDA). For a brief official history of the group see CGIAR, *1985 Annual Report*, pp. 1-14.

26. In 1971 the *Plant Introduction Newsletter* became the *Plant Genetic Resources Newsletter*. See Wilkes, 1983, for some of the historical developments in international genetic resource conservation efforts.

27. Mooney, 1983, pp. 65-9.

28. Frankel, 1987, p. 25. Frankel was a key figure in the formation of IBPGR.

29. Hawkes, 1985.

30. FAO, 1981a, pp. 56-9.

31. Plucknett et al., 1987, p. 143.

32. The case of IR-36 is outlined in ibid., pp. 171-85.

33. Alvares, 1986.

34. One is reminded of Whitehead's view of the great thinkers who consolidated the modern scientific thought. 'Their triumph was overwhelming: whatever did not fit into their scheme was ignored, derided, disbelieved', Whitehead, 1926, p. 74.

35. Chang, 1984, pp. 251-3.

36. Richharia was retired from CRRI because of this conflict with IRRI. The strength of institutional pressure and linkages between global research network, donor agencies, foundations and determined scientists led to the fall of one of India's best rice breeders. He sought refuge in the state of Madhya Pradesh. 'Here, this . . . gifted and imaginative rice scientist maintained over 19,000 varieties of rice *in situ* on a shoestring budget of Rs 20,000 per annum, with not even a microscope in his office-cum-laboratory . . . His assistants included two agricultural graduates and six village level workers, the latter drawing a salary of Rs 250 per month. Richharia had created, practically out of nothing, one of the most extraordinary living gene banks in the world . . .,' Alvares, 1986, p. 11.

37. Ibid., p. 11.

38. The removal of Richharia made it easier for Green Revolution advocates to rise to higher positions in the Indian research establishment. Similar patterns of removing those who favour particular lines of research was practised in the US early this century in order to promote the use of hybrid corn.

39. Hanson et al., 1984, p. 1.

40. See Mooney, 1983, pp. 46-52. Mooney argues that plants have higher commercial value than cultural artefacts and therefore should be subject to repatriation to the countries of origin.

41. Cunningham, 1987, p. 13.

42. The Plant Genetic Resources Centre at Addis Ababa hold over 1,000 lines of teff.

43. Perdue et al., 1986, p. 65. More recent efforts to grow *Vernonia* for export were undertaken at Mombasa, Kenya. Attempts to grow the plant in Jamaica failed but plants introduced in Puerto Rico show promising results.

44. Mooney, 1983, p. 75.

45. US Congress, *Department of Agriculture Can Minimize Crop Failures*. See also Sun, 1986a.

46. IBPGR, 1987.

47. RAFI, 1987d, p. 2.

48. Ibid., p. 4.

49. Myers, 1979, p. 61.

50. 'The problem arose because of the cytoplasmic uniformity of a large proportion of the maize being grown at that time, and, for the first time, breeders came to realize that disease susceptibility is not determined solely by nuclear genes', Brown W. L., 1983, p. 5.

51. See NAS, 1972. In 1970, six cultivars of cotton represented 68 per cent of the land covered by cotton in the US, six cultivars of soybean covered 56 per cent and six cultivars accounted for 41 per cent of the cultivated wheat.

52. Duvick, 1984, p. 164. See also Brown W. L., 1983.

53. Day, 1987, p. 40.

54. Doyle, 1985, pp. 13-14.

55. These topics were major themes in the 1980s and will not be analysed here except in so far as they relate directly to the question of genetic resources.

56. Sharma, 1987, p. 14.

4. Branching Points in Biotechnology

The evolution of conventional plant breeding methods in conjunction with advances in chemical technology helped to shape the current picture of world agriculture, but this model is already reaching limits. To move agriculture into alternative patterns of production required major changes in technology. Plant breeding enabled the industry to make relatively minor modifications in plants to achieve desired objectives such as high fertilizer responsiveness. Following the unravelling of the genetic code, scientists have been able to make major advances in their understanding of genetics. Genetic engineering, tissue culture, cell fusion and other techniques have opened up immense opportunities for producing new plants and animals. These technological possibilities have introduced significant discontinuities and convergences in both industrial and agricultural production.

Like micro-electronics, biotechnology has the potential to reorganize large sections of the industrial and agricultural sector. The introduction of new life forms into the economic sphere is likely to lead to major economic changes and, while it is still early to comprehend the full scale of these changes some trends are already starting to emerge. The ways in which these changes are likely to affect Third World countries will dwarf the impacts of the Green Revolution; furthermore, these innovations are likely to be applied in ways similar to the Green Revolution. This negative picture is only part of the story. The fact that biotechnology is science-intensive makes it possible for Third World countries to acquire a certain measure of technological capability through long-term education and training programmes. Acquiring such capability may contribute to renewed growth in agriculture and industry in Third World countries.

Biotechnology industry

The historical evolution of biotechnologies is characterized by three major phases or generations.[1] The first generation of biotechnologies was based on empirical practice with minimal scientific or technological inputs. This generation dates as far back as 7000 BC and its products include the traditional methods of fermentation to produce food and drink. The historical development of disciplines such as biochemistry is associated with efforts to understand the process of fermentation.[2] The institutional requirements were minimal as well since most of the production was done at family level. The knowledge acquired during this phase was considered to be communal and was passed on to subsequent generations as part of the cultural heritage. This process was partly to ensure that the utilization and preservation of the knowledge was diversified. The history of industrialization is partly a story of the separation of production knowledge from its end-use (products).

The growth of industry, especially in the 1930s and 1940s, was associated with the increased application of science to large-scale microbiological processes; this ushered in the second generation biotechnologies. A combination of advances in biochemistry and chemical engineering contributed to new methods of fermentation, bioconversion and biocatalysis, which led to numerous products in the pharmaceutical, chemical, fuel, and food industries.[3] Major advances in these areas resulted from a long history of institutional reorganization, especially in Germany which pioneered in the application of scientific knowledge to biochemical processing. Second generation biotechnologies relied heavily on the use of identifiable natural processes and was largely based on the selection of particular biological processes and their enhancement for industrial production. These innovations contributed immensely to industrial output over the post-World War II period. Technologies have continued to improve and new lines of innovations now include enzyme immobilization and plant tissues propagation.

Major changes that occurred in the biological sciences and solid state physics in the 1960s and 1970s led to revolutionary development in biotechnology. The ability to manipulate the genetic code enabled researchers to produce novel genetic combinations for industrial applications. The use of recombinant DNA (rDNA) and hybridoma (cell fusion) techniques made it possible to modify life forms for a wide range of industrial applications. These techniques constitute third generation technologies. It is important to note the role of other technologies in the advancement of biotechnology. Of particular importance is micro-electronics whose application in microbiological research played a major role in improving the quality of experiments, data collection and analysis as well as performing simulations.

These generations of biotechnologies have continued to co-exist, which suggests that technological innovation is not a linear process that would be characterized by the mechanistic neo-Darwinian framework under which the market works as a set of selection mechanisms: the suitable technologies being selected while the unsuitable are consigned to the archives of patent history.[4] Instead, there are multiple paths pursued by the technologies as they articulate themselves in the economic system. While some industrialized countries are applying the state of the art technologies to areas such as fermentation, Third World countries are using other ways of improving yields in the same processes. This suggests that the process of technological innovation drifts towards complexity and diversity. The prevailing historical conditions in the various countries are therefore a major factor in influencing the paths of technological development in biotechnology.

Biotechnology as an industry has its origins in university research in microbiology and genetic engineering. Most of the firms set up in the US in the 1970s and early 1980s were composed of university researchers who had moved into the corporate world with the scientific knowledge and research tools they had acquired while working at universities. This is, by definition, a science-intensive industry; a feature that makes it relatively different from other industries which are capital-intensive and involve the use of extensive mechanical equipment. More than US$4 billion had been invested in the US biotechnology industry by 1985, of which 75 per cent was in the pharmaceutical sector (43 per cent for cancer research, 19 per cent for other therapeutics and 13 per cent for diagnostics).[5] Some US$479 million had been invested in plant improvement and US$154 million in agrochemicals, accounting for 12 and 4.0 per cent respectively. Nearly 62 per cent of the investment had gone into recombinant DNA (US$2.47 billion) and 30 per cent (US$1.2 billion) was in hybridomas and monoclonal antibodies. About 8.0 per cent of the amount had been invested in fermentation and tissue culture research.[6]

Biotechnology research started showing potential products in its early stages; the watershed period was 1978. Harvard University researchers produced rat insulin, and Genentech (San Francisco), one of the earliest biotechnology firms, produced human insulin. The same year researchers at Cornell University implanted into a yeast a gene that coded for the production of the leucine amino acid. The following year Genentech produced human growth hormone and, in 1980, two kinds of interferon. Biogen of Switzerland put Europe on the map by producing human interferon. These advances, in addition to the decision of the US Supreme Court allowing for genetically engineered micro-organisms to be patented, transformed the image of biotechnology in the marketplace. When Genentech offered its stocks for sale in October 1980, Wall Street went wild. The stock rose from US$35 to US$89 per share within minutes and within hours of frantic trading the firm raised US$55 million. For a company that did not have a product on the market, the sale was quite astounding. In 1981 Cetus raised a record US$115 million in the first stock offering.[7]

Table 4.1
Funding for U.S. biotechnology industry, 1985 (US$ million)

Source	Amount	Percentage
Equity	2,581.0	65.0
Contracts and joint ventures	578.0	15.0
R&D Limited Partnerships	55.0	14.0
Grants to universities	260.0	6.0
Product licences	14.0	Negligible
	3,991.0	**100.0**

Source: Murray, '*The First $4 Billion*,' p. 293.

The market potential generated by a series of technological breakthroughs signalled to large firms that biotechnology offered a new source of industrial renewal. The industry has since then become increasingly concentrated because of takeovers. This is not only reflected by the number of large firms in the industry, but also by the composition of the main carriers of technology as shown by the total investment. About 63 per cent of funds (US$2.6 billion) is accounted for by ten organizations. The main source of funds for the industry in the US is venture capital, while European firms have relied heavily on government support. It is becoming difficult to obtain venture capital for industry and the firms are desperately trying to put new products on the market so that they can attract new funding.

Table 4.2
Number of firms involved in biotechnology, 1985

Area	France	Italy	Germany	UK	USA	Japan
Agriculture	5.0	1.0	2.0	15.0	73.0	12.0
Antibiotics	1.0	2.0	4.0	1.0	4.0	8.0
Chemicals	1.0	—	1.0	4.0	37.0	31.0
Diagnostics	3.0	5.0	6.0	10.0	141.0	15.0
Fermentation	3.0	—	—	6.0	21.0	13.0
Food	2.0	—	1.0	12.0	18.0	17.0
Hybridomas	2.0	4.0	4.0	4.0	50.0	13.0
Pharmaceuticals	2.0	5.0	4.0	5.0	28.0	28.0
Total	19.0	17.0	22.0	57.0	372.0	137.0

Source: Dibner, '*Biotechnology in Europe*,' p. 1368.

Although Japan has a long-term strategy on biotechnology, their total investment in the industry is still low. By 1985, the country had invested US$23 million in biotechnology of which 85 per cent was in the health sector. Only about 10 per cent of the investment was in the agricultural sector (crop improvement and agrochemicals). About 65 per cent (US$15 million) of the funding came from contracts and joint ventures while 35 per cent was from equity participation. Japan had devoted 80 per cent of the allocation to rDNA techniques while 12.5 per cent went to hybridoma research. Only about 5.0 per cent was invested in tissues culture technology.

Because of the reduction in venture capital in the US in recent years, and the uncertainty associated with biotechnology, new forms of research funding have evolved. One of them is investment in specific lines of research as limited partnerships. This enables the investor to risk funds only on a particular product and not invest in the whole firm through equity participation. Biotechnology firms have also undertaken contract work for large firms which allows large firms to maintain flexibility in their investment strategies while they decide on which areas to build in-house biotechnology capability. The advances in the US biotechnology industry were associated with new forms of institutional organization that helped support the process of innovation.

Table 4.3
Leading US biotechnology firms, 1985 (US$ Millions)

Firm	Turnover (US$ Million)	Net Profit
Genentech	89.6	5.6
Cetus	54.9	1.4
Biogen	31.4	−19.1
Centocor	22.4	3.5
Amgen	19.8	−1.5
Genex	16.2	−15.9
California Biotech	9.6	−0.5
Collaborative Research	8.8	4.3
Molecular Genetics	8.3	−2.5
Integrated Genetics	7.3	−3.7

Source: Shearson Lehmann Brothers Inc., quoted in *The Economist*, 'Biotechnology's Hype and Hubris', p. 93.

While Third World countries are attempting to reduce their dependence on imported agricultural inputs, some biotechnology research is being directed at increasing the consumption of agrochemicals. Biotechnology researchers have a wide range of options for dealing with disease, pests and weeds. One is to produce disease-resistant varieties, which is the conventional approach among breeders. Such breeding activities have far-reaching implications for the US$30-billion dollar pesticide and fertilizer industries. The corporate response to this challenge is to enter the biotechnology business and breed for characteristics that do not undermine agrochemical sales. Over 28 firms had launched 65 research programmes in herbicide tolerance by 1987. The total market value of introducing herbicide resistant crops is expected to reach US$6.0 billion by the year 2000. The seven leading world pesticide producers, which account for US$10.9 billion or 63 per cent of the global pesticide sales, now have research projects on herbicide tolerance. These companies are also leading seed producers.[8]

Table 4.4
Leading pesticide producers

Firm	State	Pesticide Sales (US$ million)	Global Percentage	Herbicide Tolerance Research
Bayer	West Germany	2,344	13	Yes
Ciba-Geigy	UK	2,070	12	Yes
ICI	UK	1,900	11	Yes
Rhone-Poulenc	France	1,500	9	Yes
Monsanto	USA	1,152	7	Yes
Hoechst	West Germany	1,022	6	Yes
Du Pont	USA	1,000	6	Yes
TOTAL		10,988	63	

Source: Rural Advancement Fund International, Pittsboro, North Carolina, USA.

The US agriculture pours some 0.5 billion tonnes of toxic active ingredients on farmland annually, with only 1.0 per cent reaching its target. In addition to this low level of efficiency, about 30 species of weeds and 447 insect species have developed resistance to agrochemicals designed to affect them. It is estimated that errors in applying herbicides to US maize, soybean and wheat cost farmers US$4.0 billion a year. With the cost of developing new plant varieties being US$2.0 million and that of new pesticides exceeding US$40 million, the industry finds it more logical to adapt plants to agrochemicals than chemicals to plants. Developing chemical-adaptable plants is expected to raise the sales of agrochemicals and work is already underway to bring such plants in production in the next few years.[9]

For example, Monsanto has been working on breeding herbicide-resistant plants which enable farmers to use more herbicides without affecting the plants.[10] Herbicide resistance already exists in nature. It is through the existence of natural resistance that selective herbicides are made applicable, otherwise all plants, including crops, would be destroyed on application. Wheat, for example, is resistant to Glean, a sulfonylurea compound produced by Du Pont.[11] One of Monsanto's research projects is to develop soybean varieties resistant to Roundup, a wide spectrum herbicide it produces. Such a variety would help increase the sale of Roundup by Monsanto by some US$150 million yearly. The firm has also to compete with other biotechnology companies that might produce disease-resistant varieties. Monsanto has been trying to transfer a gene that codes for pest-resistance from a bacterium into the tomato. Such a tomato would produce its own insecticide. The first such tomatoes were planted in June 1986 in Illinois, USA.

In June 1987, Calgene received the approval of the US Department of Agriculture (USDA) for field trials with genetically modified tobacco plants containing its GlyphoTol herbicide tolerance gene. The gene was successfully expressed in tobacco plants in February 1985 and patented six months later. This gene protects sensitive crops against Monsanto's glyphosate, one of the most effective herbicides available on the world market. Calgene also started field trials with phenmedipham tolerant Canola plants in Manitoba, Canada, containing its BromoTol herbicide tolerance gene which was successfully cloned in February 1987; the gene codes for an enzyme that detoxifies the herbicide bromoxynil. It was cloned from a naturally occurring soil bacterium. Earlier greenhouse tests with plants containing the gene were tolerant at levels seven times the normal field rates.

A recent development in this field has been the work of the Belgian firm, Plant Genetic Systems. This firm has developed an engineered gene with an enzyme that inactivates the active ingredient of Hoechst's Basta herbicide. The gene will be inserted in potato, tobacco and tomato plants and enable more Basta herbicides to be used without harming the crops. Trials have shown that plants containing the gene can withstand up to ten times the doses normally used for weed killing. The application of the gene is expected to earn Hoechst another US$200 million a year. Such research makes sense from a corporate point of view but does not necessarily reflect the long-term interest of Third World countries.

Already, the largest corn-seed producer in the world, Pioneer Hi-Bred has been given an altered gene for herbicide resistance developed by American Cyanamid. The offer allows Pioneer to incorporate the gene in their corn seeds and therefore increase the use of imidizolinone family of herbicides developed by American Cyanamid. It is not known how far the collaboration between the two firms is likely to go. Pioneer Hi-Bred is predominantly a seed firm which is starting to build capability in other fields. The future of the collaboration depends largely on the corporate arrangements entered by the two firms. It is expected, though, that seed firms are going to continue to acquire capability in other fields while agrochemical firms are going to increase their stake in the seed industry.

This is part of the restructuring process that is underway and it is still difficult to tell to what extent seed specialist firms are likely to remain in the market without adopting the new strategies of linking agrochemicals to seed production.

Table 4.5
Research in herbicide-resistant crops: a sample

Herbicide Producer	Contractor	Crop	Resistant To
American Cyanamid	Phyto-Dynamics	Maize	Prowl
American Cyanamid	Molecular Genetics	Maize	Imidizolinones
American Cyanamid	Pioneer Hi-Bred	Maize	Several
Eli-Lilly	Phyto-Dynamics	Maize	Treflan
Monsanto	Phyto-Dynamics	Maize	Roundup
Monsanto	In-house	Several	Roundup
—	Calgene	Several	Roundup
Kemira Oy	Calgene	Turnip-rape	Several
Kemira Oy	Phytogen	Cotton, soybean, tobacco, tomato	Several
Rhone-Poulenc	Calgene, in-house	Sunflower	Bromoxinyl
Ciba-Geigy	In-house	Several	Atrazine
Shell	In-house	Maize	Cinch
Shell	In-house	Several	Roundup
Dekalb-Pfizer	Calgene	Maize	Not specified
Lubrizol	Phyto-Dynamics	Oil seeds	Not specified
Hoechst	Plant Genetic Systems	Tomato, tobacco, potato	Basta
Du Pont	In-house	Tobacco	Glean
Monsanto	In-house	Sorghum	Bronco

Source: Hobbelink, *New Hope or False Promise*; Rural Development Fund International, Pittsboro, USA

The rise of small biotechnology firms poses a major threat to the seed companies, especially those specializing in seeds that are relatively easy to genetically engineer. One of the strategies of the emergent biotechnology firms is to acquire firms that are breeding crops which they have the capacity to manipulate genetically. This trend has already started. Calgene has recently acquired Stoneville Pedigree Seed which specializes in cotton seed. The acquisition of the firm enabled Calgene to record its first profitable quarter in May 1987 (US$493,000 on revenues of US$7,353,000). Calgene acquired Stoneville because it has patented the GlyphoTol herbicide tolerance gene for cotton and needs to have control over the breeding of the crop. Since the genetically-engineered plants are resistant to a Monsanto herbicide, it is likely that Monsanto will increase its interest in cotton breeding.

Recent advances in plant genetic engineering are opening numerous avenues for innovation. With the help of rDNA technology, scientists are now able to transfer genetic material from micro-organisms to plants to perform particular functions. An example is the work on *Bacillus thuringiensis*. The bacterium produces a group of proteins that are lethal to most moth and butterfly, or lepidopteran, larvae. The toxin from this bacterium is usually sold as an emulsion for spraying on crops. Plant scientists have succeeded in isolating the gene coding for the toxin protein and this gene can now be used in at least three different ways.

First, with knowledge on the functioning of the gene, it is possible to improve the efficiency of the micro-organism. Second, the gene can be transferred into other micro-organisms associated with vulnerable parts of crops, such as roots. Monsanto scientists have put *B. thuringiensis* (subspecies *kurstaki*) toxin genes into *Pseudomonas fluorescens,* a micro-organism associated with the roots of corn plants. This enables the toxins to affect the corn rootworm. A third approach is to insert the genes into the crop so that it can generate its own toxins. This enables the toxins to be used against pests that attack those parts of the crop that cannot be reached by spraying, such as roots or cotton bolls. Scientists are actively searching for fungal or bacterial sources of insecticides which can be used like *B. thuringiensis.*[12]

Another focus of plant biotechnology is nitrogen fixation, a process that enables some bacteria to provide plants with nitrogen through a symbiotic relationship. Current research has concentrated on the *Rhizobia* bacteria. With genes from such bacteria, and others such as *Klebsiella pneumoniae,* scientists are working on the possibility of enabling crops to produce their own nitrogen and therefore reduce fertilizer costs. Instead of modifying plants, other firms are looking into ways of modifying the environment to favour higher crop productivity. Advanced Genetics Sciences (USA) has been working on ways of reducing crop loss through frost, which is estimated to cost over US$1.0 billion in the US and nearly US$4.0 billion worldwide. Frost is caused by a protein in the *Pseudomonas syringae* which induces ice crystals to form when the temperature drops just below freezing point. The approach taken by Advanced Genetics Sciences is to delete the gene that codes for the protein in the bacteria and then release them into the environment. The 'ice minus' bacteria have already been released for testing in the US.[13]

Animal production has been one of the areas of biotechnology research. The range of research covers animal genetic engineering, animal reproduction, growth and development regulation, nutrition, disease and pest control and behavioural control. Already, genetically engineered bovine growth hormone (bGH) which stimulates lactation has been developed by Genentech for Monsanto. An injection of 44 milligrams per day per cow has resulted in an increase of 10-40 per cent in milk yield.[14]

The wide application of this technique is likely to transform the dairy industry irreversibly. Advances are also being made in the transfer of genetic material from one animal to another and producing 'transgenic' animals. This not only opens up the possibility of breeding new types of animals, but also using animals as factories for producing industrial products. Scientists at Integrated Genetics announced in 1987 that they were able to produce tissue-type plasminogen activator (t-PA), a heart drug, from the milk of genetically-engineered mice. The use of transgenic animals is likely to increase and will possibly replace some aspects of industrial production.

Animal biotechnology is also starting to complement or replace traditional methods such as artificial insemination by embryo transfer. This method enables scientists to stimulate superovulation in donor cattles and transfer the fertilized embryos into surrogate cows for rapid stocking. With this method breeders can increase the population of animals with genetically-required or engineered traits much faster than in the traditional methods. There are about 140 companies working on embryo transfer in the US and the market is worth US$30 million yearly. Ireland has turned the technique to improving horses. Recent changes in the US law to allow for the patenting of animals will stimulate the animal biotechnology industry. The techniques also offer the potential of large-scale reproduction of some wild animals currently threatened with extinction. Possibly countries such as Kenya could use animal biotechnology to rebuild rhino populations and other threatened animal species.

Biotechnology work is not restricted to industrialized countries. A large number of Third World countries have formulated or are preparing biotechnology policies, most of which are still statements of intent and focus on public sector research, although some Third World countries have allocated resources for the implementation of the policies. Policy statements in some Third World countries constitute the consolidation of research that is already underway, but they also identify new research areas. India was one of the first Third World countries to formulate a long-term biotechnology programme, which has identified health, industry, agriculture, veterinary and animal breeding, energy and environment and ecology as priority areas. Unlike other Third World countries, India has a large pool of scientists in these fields to enable the country to implement the plan. Other countries have incorporated biotechnology concerns in their national development strategies.

The establishment of the International Centre for Genetic Engineering and Biotechnology (ICGEB) in New Delhi and in Trieste, Italy, and the continuing activities of the United Nations Industrial Development Organization (UNIDO) are likely to facilitate the formulation of biotechnology policies in the Third World countries. ICGEB was formed in 1983 after a long international struggle between the Third World and industrialized countries. This was partly because ICGEB was set up to deal mainly with the problems of Third World countries, focusing on agriculture and health, which in part is why the industrialized countries have been reluctant to support the institution. ICGEB is expected to start functioning as an intergovernmental agency when a minimum of 24 countries ratify its statutes. By 1987 only 13 countries (Algeria, Bhutan, Bulgaria, Cuba, Egypt, Hungary, India, Iraq, Kuwait, Panama, Senegal, Venezuela and Yugoslavia) had done so.

Some of ICGEB's functions are to promote research, provide technical support and policy guidance to Third World countries. Like most other international organizations, ICGEB aims at operating closely with national organizations in line with national policies. The effectiveness of these policies will depend on the extent to which the countries will promote the commercial application of the innovations. ICGEB is supported by 40 countries and is expected to start working in 1988. The Trieste wing will deal mainly with vaccines, diagnostics, monoclonal antibodies, enzyme engineering, cellulose waste, lignocellulose waste, metal leaching and food fermentation. The programme for the New Delhi wing covers nitrogen fixation, trypanosomes, agricultural plants, hepatitis B vaccine, genetic transformation of *Graminaceae,* plant cell culture and propagation, plant pathogens, malaria chemotherapy, Chagas disease, malaria, schistosomiasis, gene engineering, plant cell and pollen propagation.

ICGEB's operation will depend largely on the level of co-operation with national research institutes. So far it is still trying to attract the support of the industrialized countries, especially the US which has remained firm against the establishment of an international agency that could possibly compete against US biotechnology firms or help strengthen the capacity of the Third World countries to undertake biotechnology research. The possible impact of ICGEB, beyond raising awareness on biotechnology policy issues, is likely to be minimal. The potential for stimulating research in microbiology in the Third World countries is likely to stem from less institutionalized arrangements, such as the Microbiological Resources Centres (MIRCENs).

The MIRCENs represent a unique institutional arrangement which relies largely on networking. The idea to set up the MIRCENs originated with UNEP in 1973, following a recommendation at the United Nations Conference on the Human Environment in Stockholm, calling for the conservation of genetic resources. They are administered by UNEP. UNEP conceived a network of research activities which would involve the use of microbiological processes for environmental management and rational resource utilization. So far MIRCENs have been established in Brazil, Egypt, Guatemala, Kenya, Senegal and Thailand. Their functions include exercising regional responsibility for the collection and maintenance of microbial genetic resources in reliable culture collections. The MIRCENs also serve as training and R&D centres with emphasis on technologies relevant for the Third World. Pilot projects are currently being implemented on using microbial processes to enhance soil fertility, produce energy from agro-industrial waste, degrade persistent pollutants, and use biological and vector control.[15]

The MIRCENs work closely with the World Data Centre on Microorganisms at the Institute of Physical and Chemical Research (RIKEN) in Japan. The data centre serves as a world register of information on microbial strains and is available to all users. The network has also started operating an International Microbial Strain Data Network (MSDN) at the Biotechnology Centre, Cambridge University, UK. This network, together with a joint UNIDO/ WHO/UNEP Working Group on Biotechnology Safety established in 1986, could make a major contribution in building biotechnology capability in the Third World. The MIRCENs are still using second generation biotechnologies but the setting up of ICGEB creates an ideal environment for the establishment of genetic engineering projects in the MIRCENs. Instead of ICGEB setting up its own network of researcher institutes, it may be more relevant to incorporate the MIRCENs into the new research lines pursued by the centre.[16]

One of the major functions of the state in facilitating technological development is the setting up of relevant institutions. Countries such as India, Brazil and Mexico have set up institutions devoted to the promotion of biotechnology research. The linkages of these institutions vary according to the policies adopted by the state towards the private sector. India has earmarked US$400 million for biotechnology over the 1985-90 period. The programme will be co-ordinated by the National Biotechnology Board created in 1982. Postgraduate biotechnology programmes have been initiated in a dozen universities as part of the board's training efforts. The board plans to build animal-breeding and embryo transfer centres, genome laboratories, gene banks, and biotechnology manufacturing units.

Mexico has recently devoted large amounts of money to institution building. In 1985, in addition to an existing centre for research in nitrogen fixation, Mexico opened the Centre for Genetic Engineering and Biotechnology at Cuernavaca. The US$3.0-million centre will be involved in the production of microbial insulin. Although there is a large biotechnology research community in Mexico, the country has not established strong links between the private sector and research community. As a result, the research findings are not effectively translated into commercially viable products.

Other countries that have launched biotechnology programmes include

Indonesia and Thailand. The government of Indonesia has invested more than US$30 million in a new Institute of Molecular and Cell Biology. The investment is part of a long-term programme to provide skilled manpower, which has been the main obstacle to the government's biotechnology initiatives. Much of the initial training will be devoted to disease researchers and cell biologists. The institute is aimed at providing a base for co-operation between the research establishment and the private sector.[17] Thailand has also built a National Centre for Genetic Engineering and Biotechnology under the Ministry of Science, Technology and Energy, its functions include strengthening the country's capability in genetic engineering and biotechnology, industry-university links, and the co-ordination of government and international support for local institutions in biotechnology and genetic engineering.

Although the African countries are reshaping their economic policies to reflect the needs of the agricultural sector, there are still no major efforts to formulate biotechnology policies. It is instructive to note that the UN General Assembly special session on the African crisis, held in May 1986, did not lay any specific emphasis on the application of science and technology to development. Most of the major policy requirements still relied on financial and monetary reforms. The need to emphasize the potential role of biotechnology was obviously lacking in the preparatory papers and the deliberations.

Some African countries have not yet consolidated their agricultural research activities. So long as these countries do not recognize the linkages between innovation and institutional reorganization, they are not likely to formulate viable biotechnology policies. Despite this, there are institutions working on various aspects of biotechnology. These include government institutions as well as inter-governmental organizations such as the Kenya-based International Centre for Insect Physiology and Ecology (ICIPE). In addition, these countries have a relatively large pool of skilled people working in biotechology-related fields.

Some of the activities of the private sector in the field of biotechnology are subsumed in long-term research agreements signed between the Third World and industrialized countries. Although these agreements appear as public sector initiatives, their implementation involves the participation of the private sectors in ways that are difficult to establish in advance. In 1985, India and the US signed a science and technology agreement in which biotechnology issues featured prominently. The Gandhi-Reagan Science and Technology Initiative (GRSTI) focuses on applied biology, genetic engineering and plant biotechnology. Under the agreement, the two countries will collaborate in developing recombinant vaccines and drugs for measles, cholera, typhoid, malaria, non-A/non-B hepatitis, rabies and leprosy. Collaboration will also be extended to the development of a male contraceptive vaccine.

State co-operation in biotechnology projects is likely to follow the traditional patterns established by various countries. Similar patterns have been observed in the area of renewable energy research. This co-operation may also take the form of international aid. But unlike the renewable energy sector, a large share of the products of biotechnology research are not specifically aimed at Third World markets. Some of the agreements have raised major questions on the nature of collaboration between the industrialized countries and the Third World. For example, it has been argued that the GRSTI has opened up the possibilities for the control of Indian biotechnology by the US. Moreover the agreement could also provide US firms with a base in which to test new biotechnology products.

In countries such as Brazil, the development of particular technologies has been achieved through closer collaboration between the private sector and public institutions. This was indeed the case in the development of the fuel alcohol programme in the 1970s. The private sector, especially the sugar industry, managed to influence the government to launch a programme that would utilize surplus sugar to produce fuel alcohol for automobiles; this was not a new venture as Brazil had a long history of fuel alcohol production. This process was relatively easy because the sugar industry was well-established and there existed a large number of trade associations representing the private sector and acting as lobbying and pressure groups. In addition, sugar-cane farmers in Brazil were already organized under powerful co-operatives that influenced government policies on matters ranging from research to pricing.

The case of biotechnology in Brazil is slightly different, although it manifests the same pattern of forging closer ties between the private and public sectors. Unlike sugar, there is no homogeneous biotechnology body in the country, although some of the main actors in the biotechnology field are associated with the sugar industry. The impetus for institutional reorganization was started in 1985 when the government established a Secretariat for Biotechnology. The government also allocated US$52 million for a technology programme that included biotechnology as a priority and, through the Empresa Brasileira de Pesquisa Agropecuaria (EMBRAPA), has set up the Centro de recursos Genéticos (CENARGEN). The centre is involved in germplasm preservation, genetic engineering, tissue culture, and biological pest control research.

There is a growing link between universities and industry as university researchers prepare to adapt their work to commercial production. The private sector is also increasing its involvement in biotechnology. Firms such as Biomatrix and Bioplanta are increasing their activities in the supply of seed to farmers. Some of their new products include virus-free strawberry and potato seed sold to farmers, and these firms aim at displacing foreign seed companies from the local market. In 1986, eight Brazilian firms involved in biotechnology-related activities formed the Associacão Brasileira des Empresas de Biotecnologia (ABEB), which is currently working on a national biotechnology policy, technology exchange and the development of specialized expertise in various fields. ABEB is working closely with the government through EMBRAPA.

The process of technological change does not necessarily rely on the linkages between the private and public sectors. Although the incentives provided by the market serve to stimulate technological innovation, it is possible to arrange institutions in such a way that they promote technological change without relying on market incentives. This case is illustrated by the efforts undertaken in Cuba to promote biotechnology research. For example Cuba's interferon research is conducted largely at the Centre for Biological Research (CIB) in Havana. Biotechnology research in Cuba reflects the country's economic situation; it has opted for biotechnology because it is research-intensive rather than capital-intensive.[18] The low capital outlays required for biotechnology research give countries with a diverse scientific community relative advantage in pursuing particular research lines. What is significant in such a situation is the technology policies adopted to promote innovation.

Cuba has opted for a strategy similar to that adopted by Japan during its 'catch-up' period. Instead of focusing on the state-of-the-art work, Cuba has placed emphasis on applied research. The interferon programme, for example, has relied on international novelty as a base but has pursued research lines that seem promising. As a result, Cuba has been able to bring itself close to the R&D frontier. The programme relies heavily on the international acquisition of knowledge; the methods used include foreign training of Cuban scientists, invitations to foreign scientists, acquisition of publications, and the widespread diffusion of knowledge. The Cubans chose interferon production partly because there was abundant knowledge on the subject and its feasibility was proven.

The success of the programme is illustrated by the fact that Cuban scientists have been making findings similar to those reached in the industrialized countries. So far interferon production laboratories have enough buffy coat to produce 4.0 milligrams a day. As a result of the success of the programme Bioquimica do Brasil (Biobras) of Brazil has signed a contract to exchange Cuban interferon technology for Brazilian microbial production insulin technology. Brazil intends to apply the Cuban approach to other sections of the economy; the interferon model was partly a process of institutional learning. In addition, the scientists have developed novel ways of DNA synthesis and enzyme restriction. This is partly in response to the embargo placed on the country by the US, which is the main source of biotechnology materials and equipment.

There is a growing number of joint ventures between firms and institutes in industrialized countries and the Third World. These ventures cover various aspects of biotechnology ranging from animal to plant processes. For example, the Chinese government has entered into a joint venture with International Embryos to boost the production of dairy cattle using embryo transfer techniques; this venture will be based at the Jinah International Embryos Centre in Guang Zhou province. The centre's aim is to increase the number of dairy cattle in the province from 40,000 to 0.5 million in five years.

Other joint ventures involve the use of surplus local resources for industrial application. The Institut Français de Recherche Scientifique pour le Développement et Coopération in Paris has signed a joint venture with the Costa Rican Centro de Investigaciones de Tecnología Alimentaria (CITA) to turn waste bananas into high-value animal feed. The country dumps 140,000 tonnes of sub-standard bananas into the ocean every year. The joint venture will utilize a solid-substrate technology to ferment the banana (pre-dried to 10 per cent water content). The process will use the fungi *Aspergillus niger* (which has a high amylase activity) to turn the banana substrate into flour with 20 per cent protein content. The process was first tried in Martinique but discontinued because of irregular and expensive power supply as well as limited banana availability.

Other recent joint ventures include an arrangement between China's Centre for Biotechnology Development and Promega Biotec of Wisconsin (US) to produce restriction enzymes in Luoyang, Henan Province, which will enable China to build capability in the field of enzyme restriction production for a wide range of applications.[19] This is an interesting venture because it is entered into between the government and a foreign firm, a factor that signifies the enabling role of the state in the early stages of technological evolution.

One of the most significant developments in the evolution of biotechnology is the formation of joint ventures and collaborative programmes between the Third World countries. It is notable that the collaboration is starting at an early stage of the development of the industry. This reflects both the changing policy regimes of the Third World countries as well as the techno-economic requirements for international collaboration. In Brazil, both the government and the private sector have been active in the search for collaboration with other Third World countries in agriculture and medicine. Biobras has earmarked some US$2.0 million to develop gene-spliced interferon and insulin in collaboration with Instituto Sidus ICSA and Polychaco SAI of Argentina.

The co-operation between Third World countries has been not only in the field of marketing but has also been aimed at exchanging technological knowledge. This is also the case in the agreement signed between Brazil and Cuba for the exchange of insulin and interferon know-how. These trends are likely to continue, especially given the closer co-operation emerging among some Third World countries. African countries, however, do not seem to be seeking research opportunities for South-South co-operation and still rely on project initiatives from the industrialized countries.

Third World countries' growing interest in biotechnology and the rapid changes occurring in industrialized countries is raising a large number of policy issues. While some of the countries are now formulating national policies, others are already encountering critical issues that require major policy changes. These include issues such as national policy guidelines, technological capability building, international co-operation, intellectual property and off-shore experimentations. The industrialized countries are pursuing similar lines of research and have already formulated policies that are more or less similar, partly because they have institutional arrangements that allow them to act as a group despite national differences. For example, the Organization for Economic Co-operation and Development (OECD) serves as a forum for formulating common policies among the member countries. Most African countries do not have similar institutional arrangements and therefore lack effective forums through which they can harmonize their policies on the basis of common long-term goals.

The policy scene is therefore characterized by national initiatives that do not necessarily reflect regional interests. This, however, is a necessary development which will be followed by the formation of regional programmes. Already, some regional economic groupings, such as the Caribbean, are discussing the possibility of closer regional co-operation in matters of biotechnology. National initiatives are likely to continue because of the divergences in the Third World countries. One of the areas receiving increasing policy concern is intellectual property protection. Most of the literature deals mainly with the impact on Third World countries of the current development in intellectual property protection in the industrial countries. There are, however, major changes occurring in the developing countries in this field as a result of the need to strengthen national biotechnology capability.

Institutional issues

The development of biotechnology in general, and some aspects of plant research, have been accompanied by major changes in the relationship between corporations and universities; as already indicated, biotechnology emerged largely from university research. Over the years, large corporations have formed a wide range of institutional arrangements with universities to ensure that the process of innovation is continued while the firms have access to the scientific findings. In numerous cases, universities have received large contracts from corporations under which the researchers are allowed to patent their inventions while the donor corporation retains the right to license the resulting technologies.

Table 4.6
Status of selected biotechnology products, 1987

Product	US Regulatory status	Firm/Collaborator	Indication
Alfa interferon	Marketed 1986	Genentech, Biogen Hoffman-LaRoche Schring Plough	Hairy cell leukaemia
Beta interferon	Human clinical testing	Cetus, Shell Oil	Anticancer, anti-infective
Concensus interferon	Human clinical testing	Amgen	Anticancer, anti-infective
Erythropoietin	Human clinical testing	Amgen, Kirin Johnson and Johnson	Chronic anaemia
Gamma interferon	Human clinical testing	Biogen, Shionogi Amgen, Genentech Boeringer Ingelhein, Daiichi-Seiyaku, Toray Industries, Schering Plough-Suntoray	Anticancer, anti-infective, anti-arthritic
Human growth hormone	Marketed 1985 Marketed 1986	Genentech, Kabi Vitrum Eli Lilly	Growth hormone deficiency
Interleukin 2	Human clinical testing	Cetus, Biogen, Takeda, Amgen	Anticancer
Tissue plasminogen	Market approval pending	Genentech, Boeringer Ingelhein, Mitsubishi Chemical, Kyowa Hakko	Thrombolytic agent
Tumor necrosis factor	Human clinical testing	Genentech, Cetus, Asahi	Anti-tumour agent
Hepatitis B vaccine	Marketed 1986	Chiron, Merck	Immune prophylaxis
Human insulin	Market 1982	Genentech, Eli Lilly	Diabetes

Source: Rathmann, 'An Industry View of Biotechnology'.

This growing relationship between universities and corporations has led to trends with long-term implications for Third World countries, including the direction of research, public accountability, intellectual secrecy and creativity. It is notable that most of the biotechnology research in industrialized countries is focused on specific problems in those countries or on products with high market potential. Biotechnology allows researchers to tune their techniques and application to specific problems using local materials; this is illustrated by recent research on fuel alcohol production.

Some of the R&D projects in the hydrolysis of raw materials for alcohol production have focused on tuning organisms to optimize the production of enzymes that work on specific feedstock or raw materials. A process developed in the early 1980s by the New York University, for example, was designed to optimize the hydrolysis of cellulosic materials indigenous to the state of New York.[20] In the long run, such optimizations in design will raise questions on the appropriateness of the technology to the Third World countries. With such developments, the knowledge base becomes more localized and therefore difficult to transfer to other countries.

These detailed aspects of the direction of R&D are compounded by the fact that most of the funding is devoted to areas which are important, but not priorities, for most Third World countries. Nearly 75 per cent of the US 1985 biotechnology expenditure went to cancer research, other therapeutics and diagnostics. Only about 16 per cent was devoted to plant improvement and agrochemicals. Some of the relevant research directions in the medical field carried economic biases as well. For example, research on malaria vaccines in some firms and institutes is aimed at meeting the high-income markets such as tourists and military personnel visiting the tropics, instead of being directed at the general Third World population.

The growing links between industry and universities is raising a number of policy questions. The decisions of some researchers at US universities to pursue particular research lines are motivated largely by the availability of corporate funding. A recent survey conducted in the US showed that the choice of research lines was influenced by corporate funding and there was a shift in focus from basic science to applied research;[21] this has several implications for Third World countries. First, due to its generic nature, basic scientific knowledge is more likely to be transferable than applied research findings. Second, the prospects of sending students to acquire basic knowledge would be diminished if research were redirected towards applied work, except in areas where the subject matter is relevant for the Third World countries.

Another major concern in the increased corporate support for university research has been the restriction of research findings as trade secrets. The survey showed that a biotechnology faculty with industry support was four times more likely than another biotechnology faculty to report that trade secrets had resulted from their research. Some 24 per cent of the respondents in the survey reported that their work could not be published without the consent of the financial supporter. This study also showed that there was a decline in the exchange of information among colleagues as corporate rivalry encroached upon the campus.

Table 4.7
Risk reported by biotechnology faculty

Question	'To Some Extent or to Great Extent' (%)	
	Industry Support	Non-industry Support
To what extent does industry research support pose the risk of:		
Shifting too much emphasis to applied research	70	78*
Creating pressure on faculty to spend too much time on commercial activities	68	82§
Undermining intellectual exchange and cooperation activities within departments	44	68§
Creating conflict between faculty who support and oppose such activities	43	61§
Creating unreasonable delay in the publication of research findings	40	53§
Reducing the supply of talented university teachers	40	51*
Altering standards for promotion or tenure	27	41§

* Significantly different from faculty with industry support ($P < 0.05$).
§ Significantly different from faculty with industry support ($P < 0.01$).

Source: Blumenthal et al., 'University-Industry Research,' p. 1365.

The growing secrecy surrounding corporate-funded projects may also influence the admission of foreign students to some departments, especially if the students come from countries that are deemed technologically competitive. This may be compounded by the fact that some of the Third World countries cannot guarantee patent protection for biological innovations arising in the industrialized countries. As a result, restrictions are likely to be imposed on access to vital information by students from such countries. Researchers from the Third World countries may be restricted from using their own inventions in their countries if such inventions are patented by other companies. This is applied to researchers who work in foreign countries and later wish to return to their own countries. A case in point is of a 'Mexican scientist who returned from a European laboratory in which he had developed some transformation systems. Now the Mexican scientist is being forced to bargain with representatives of his former laboratory to use the materials in Mexico.'[22]

The issue of public accountability has also been raised in relation to the increasing ties between industry and universities. The direction of accountability is likely to shift toward corporations and their research agenda. The collective ethic of the researchers may be influenced by their financial ties with industry and this may affect academic practices such as peer review as well as their views on public policy issues. The opinions of such researchers on matters that affect the Third World countries would be likely to be influenced by financial interests and the related research lines. As a result, they may not be trusted to give non-partisan advice to Third World countries as consultants or commentators on crucial aspects of biotechnology.

The growing degree of secrecy and fragmentation of the sciences may also lead to a reduction of creativity, especially due to the decline in communication between university researchers. Communication is crucial to the process of creative thinking because it allows for the recombination of existing knowledge from different sources. Communication also enables scientists to discover new metaphors that can help them identify new lines of enquiry.[23] Both fragmentation and reduced communication tend to reduce the chances of discovering new metaphors and may therefore retard the process of innovation. The freedom of creative thought that is often associated with untied university research may also be inhibited by the pressure to produce marketable products. The current institutional trends in biotechnology, coupled with the current reductions in university or public sector research funding in some Western countries, are likely to have a negative impact on innovation and creativity.

Risk, time and regulation

The risks posed by biotechnology products and processes have become a major topic of concern in recent years. While industry representatives feel that too much attention is being paid to remote and negligible risks, environmental and public interest groups are becoming more concerned with the long-term effects of biotechnology products. The problem of risk assessment is complicated by the fact that most methods used to determine the level of risk are static, deterministic and at times dangerously rational. Corporations may determine the level of risk largely on the basis of the potential effects on their profit earnings. Public interest groups and community organizations on the other hand may base their criteria on non-quantifiable criteria, such as the quality of life, environmental soundness and sustainable development objectives. Conventional risk assessment methods treat human beings as separate entities, removed from their historical context. Time is not taken as a major consideration, although it seems obvious that communities or individuals require time to understand the long-term implications of new biotechnology products before they can determine the risks involved.

Risk perception changes with the degree of uncertainty and the amount of information available to the public. The fact that most of the biotechnology innovations are governed by secrecy and intellectual property laws reduces the amount of information available to the public, a factor that increased uncertainty and reduced the capacity of the public to make reasonable decisions about the risks involved. Moreover, the risks are also subjected to change through time as other complementary technologies and changes are introduced into the economic system. It is becoming increasingly clear that one-off surveys of public opinion alone cannot be used as a basis for determining the risks involved in biotechnology products. This is partly because they may not understand the complex interlinkages between industry, universities, government officials and public health officials in the process of information generation.

A study of public perceptions of biotechnology in the US showed that 19 per cent of the people would definitely believe university statements about the risk of genetically altered organisms; only 4.0 per cent would believe the company making the product.[24] Such perceptions do not take into account the fact that university scientists are becoming increasingly linked into industry and their views on risks tend to converge with those of industry. This is also true of some federal agencies which are closely associated with the promotion of corporate interests. A clearer understanding of risk is a long-term learning process and cannot be adequately represented by snapshot surveys which do not take into consideration historical processes and the institutional reorganization of the biotechnology industry.

Ever since scientists demonstrated the capability to modify life, the issue of risk and regulation has been a major policy question. Ironically, it was US scientists who, in 1973, sought the regulation of their work. The decision by some scientists to seek regulation was based on the fear that rDNA technology could be intentionally misused. In addition, the scientists also recommended the formulation of regulation guidelines and a moratorium on research on experiments that could potentially cause cancer.[25] At that time biotechnology work was largely in the hands of researchers, and corporate interests had not taken over most of the major research lines. The situation has changed and industry is strongly fighting to reduce the degree of regulation.

Since the early concerns of scientists over biotechnology hazards, the range of applications and techniques has expanded and so have the potential risks. Although there is no evidence that biotechnology research poses particularly more serious health and environmental risks than research in other industries, the trends in research suggest that more care should be exercised. Unlike other industries, biotechnology research involves microscopic organisms which cannot be easily detected. It is also feared that the prevalent use of genetic engineering techniques will avail a large section of the population with the capacity to use modified micro-organisms for sinister purposes. Even without evoking the 'biological warfare' concerns, the intentional release of genetically manipulated organisms into the environment poses major challenges. It is, indeed, difficult to introduce regulations in a new field because of the absence of prior experience. Extensive regulation may stall research. The uncertain outcome of such research programmes, however, may have far-reaching consequences.

Table 4.8
Credibility about statements of risks[1]

How likely would you be able to believe statements about the risk of genetically altered organisms, such as bacteria that protect strawberries from frost, made by the following groups? Would you definitely believe them, be inclined to believe them, be inclined not to believe them, or definitely not believe them?

	Definitely Believe	Inclined to Believe	Inclined not to Believe	Definitely not Believed	Not Sure
University scientists	19.0%	67.0%	8.0%	3.0%	3.0%
Public health officials	15.0	67.0	12.0	4.0	2.0
Environmental groups	10.0	61.0	19.0	6.0	3.0
Federal agencies	9.0	60.0	22.0	6.0	3.0
Public interest groups	8.0	55.0	27.0	7.0	3.0
Local officials	6.0	48.0	34.0	9.0	3.0
Firm making the product	6.0	39.0	37.0	15.0	3.0
News media	4.0	39.0	37.0	16.0	4.0

[1] Percentages are presented as weighted sample estimates. The unweighted base from which the sampling variance can be calculated is 1,273.

Source: OTA, *New Developments in Biotechnology: Public Perceptions*, p. 90.

One of the basic assumptions of the Cartesian-Newtonian worldview is that scientific research can lead to the required effects if there is adequate control. Often the technology being developed is isolated from the rest of the environment. The clear demarcation of the subject makes it possible to conduct research but it also ignores the fact that the product will have to be applied in an open environment. There is a lot of uncertainty, especially in fields that involve the introduction of genetically-modified life forms. Ecologists have barely understood the complex interaction between natural organisms and their environment. The potential implications of introduction of new life forms cannot be predicted. It is notable that ecosystems are not normally in a state of equilibrium and the introduction of new life forms may lead to non-linear changes with great effects.

Take the case of herbicide-resistant plants. It is possible that the herbicide-resistant plants will hybridize with their wild relatives and therefore introduce into them the herbicide-resistance gene. If this happens, it will be difficult to control such weeds without using other herbicides. It is not likely that those developing such plants will be able to take into consideration all the possible exchanges of germplasm between crops, weeds and micro-organisms. Traits which are coded by a single gene whose allele already exists in many plants have the highest possibilities of being transferred. This potential transfer is not only restricted to herbicide resistance, but may also occur in cases of insect resistance and stress tolerance. So far there is no authoritative knowledge on the degree to which genetic material must move from one plant to another to turn it into a weed.

The available information, however, suggests that such a movement is indeed possible since 11 of the world's 18 worst weeds are grown as crops in various countries. It should be noted that some of the most notorious weeds, belonging to the *Avena, Hordeum, Helianthus, Solanum-Lycopersicum, Brassica, Raphanus, Daucus,* and *Sorghum* hybridize easily with crops. For example, Johnson grass (*Sorghum halepense*) a pernicious weed, hybridizes with commercial sorghum *(Sorghum bicolor*). This means that Johnson grass can pick up genetically engineered traits from commercial sorghum which could enhance its competitive advantage.[26] Weeds tend to thrive in the same conditions as crops. The world's worst weed, purple nutsedge (*Cyperus rotundus*) shares the same conditions as 52 different crops in 92 countries.[27] Genetic engineering may 'produce more troublesome weeds than conventional plant breeding, because the new technologies can more readily introduce and spread foreign genes into new lineages whereas selection only rearranges genes already present.'[28]

One of the risks posed by the introduction of herbicide-resistant crops is the spread of such plants as weeds; there are various ways through which such new weeds could be established. The first is by unwitting farmers who want to try out new crops; farmers are known to try out new crops without extension assistance. The second is through the collapse of agriculture due to social breakdown or disturbances; various crops are known to have turned into weeds during the last decade of political unrest in Uganda. The third route could be through accidental or intentional release. The spread of such neo-classical weeds would increase the chances of the exchange of genetic material with classical weeds, thereby endowing them with the ability to resist herbicides. The introduction of herbicide-resistant crops in regions will alert public awareness on the risks of biotechnology which could easily result in the spread of neo-classical weeds.

Genetically engineered micro-organisms may also pose unexpected risks. This is illustrated by the case of the 'ice-minus' bacteria, *Pseudomonas fluorescens* developed by Monsanto. In order to trace its presence in the soil, Monsanto scientists have introduced into the micro-organism genes from the bacterium, *Escherichia coli* (from the human intestine) which grows on lactose (the sugar found in milk). The same genes also allow the bacteria to split a chemical analogue of lactose called 'X-gal', thereby producing a bright blue-coloured chemical. These genes allow the scientists to detect the presence of *P. fluorescens* in the soil. Soils are treated with a selective medium that contains lactose as the only energy source and 'X-gal' as an indicator. Only *P. fluorescens* will form colonies in the soil, and they can then be detected because of their blue colour. The technique is so efficient that scientists can find a single bacterium in a gram of soil. This process is relatively safe so long as the bacteria remain on the intended farm. If there is run-off, they could get into water systems and subsequently contaminate dairy products. Such possibilities have not been adequately considered although it seems obvious that bacteria that break down milk sugar could pose a threat to the dairy industry.

The strict regulations in some industrialized countries on the release of genetically-engineered organisms, as well as the pressure of activist groups, has forced some firms to conduct their experiments in the Third World countries.[29] This can be done with or without the knowledge of the host countries. In most Third World countries the lack of regulation and of technical knowledge on some of the current advances as well as of public accountability on modified organisms, makes those countries more vulnerable to such experiments.

Already, US scientists have tested genetically-engineered animal vaccines in Argentina.[30] In July 1986, the Philadelphia-based Wistar Institute tested a rabies vaccine (using vaccinia as a vector) at Azul, 250 kilometres south of Buenos Aires. The station is operated by the Pan American Health Organization (PAHO) a section of the UN dealing with animal diseases. PAHO did not notify the Argentinian or US governments about the experiments. Although the experiment was deemed a success, questions were raised on the possibility of hybridization with natural microbes, especially given the fact that such a process could lead to unpredictable consequences. Following the experiment, the US issued guidelines governing the domestic release of genetically-engineered organisms; these guidelines required an approval from the government. They did not, however, give provisions for releases in other countries. This left options open for US firms to enter bilateral agreements with other countries for testing new biotechnology products.

The situation here may resemble the export of products whose use is banned or restricted in the US. But the issue involves the release of invisible micro-organisms which most of the Third World countries cannot effectively control or detect. The challenge lies on the Third World countries to monitor the emerging trends in genetic engineering and formulate guidelines to govern experiments. Most of the agreements between nations and international research stations do not require the reporting of every experiment. This, coupled with the secrecy that surrounds scientific work, creates suitable conditions for conducting such experiments without due precautions.

Legislative changes in the US in late 1986 also open up the possibility of exporting to the Third World drugs that have not been approved by the US Food and Drug Administration (USFDA). The old legislation required that if a US firm wanted to export a new drug that had not been approved by the USFDA to a country that had approved its use, it must be manufactured outside the US. The new law now allows US firms to manufacture the drugs in the US and export them to countries that have approved them for use before the USFDA approves them. The bill serves a useful function for the biotechnology industry which has a long list of drugs and vaccines awaiting approval. Given the reduction in the availability of venture capital on the US market, biotechnology firms are making all possible efforts to facilitate entry into the market with their products, especially in view of the possible challenge posed by Japanese firms.[31] These competitive strategies are likely to affect the Third World countries in ways that cannot be adequately predicted.

In the past, the strict US regulations governing the export of unapproved products have forced the firms to locate their productive facilities in Europe. This enabled the US firms to have access to European markets and the European counterparts hoped to have access to the technology. In addition, links with European firms enabled the US companies to benefit from their traditional markets in the Third World countries while at the same time having access to financial incentives provided by the European Economic Community (EEC). The incentives include loans and grants for buildings, equipment, R&D activities and assets, production expenses, and training costs.

Already, firms such as Contocor and Molecular Genetics have located facilities in Holland; Hybritech in Belgium and Damon Biotech in the United Kingdom.[32] Cetus, one of the largest US biotechnology firms, has announced that it would locate its European subsidiary in Amsterdam. The criteria they cite for the relocation include financial factors, operating conditions and external environmental conditions that influence biotechnology projects. With the reduction in venture capital into biotechnology in the US, it is expected that more firms will turn to Europe for joint ventures of subsidiary locations.[33] On the whole, the rise in biotechnology research has raised new challenges in risk management, environmental considerations and regulation. Like all evolutionary situations, the effects of the introduction of new technologies can only be established *post facto,* by which time it may be too late to do much since the effects are irreversible.

The potential impacts of introducing genetically-engineered organisms are so uncertain that given the technical possibilities available to researchers, it is necessary to look for ways that would maintain a measure of accountability over the R&D process. So far, attempts to introduce codes of conduct have been largely ineffective. Moreover, most codes of conduct have dealt largely with the commercial aspects of technology. Given the significance of modifying life forms, it seems relevant to work out a code of conduct so as to enable humanity to deal more effectively with any negative effects of biotechnology products. Such initiatives could be undertaken through UN agencies such as the UN Environment Programme (UNEP), UNIDO and the World Health Organization (WHO). Alternatively, it may be necessary to set up an international commission which will involve popular participation in the preparation of a code of conduct on biotechnology research. Technologies that can drastically alter the evolutionary potential should be subject to the highest level of public accountability. Claims that biotechnology research is safe do not necessarily discount its potential impacts. As Regal has noted, '[B]eing able to handle matches easily and safely is not a general argument for the complete safety of fire.'[34]

The last harvest?

It was noted in Chapter 1 that the emphasis placed on land, labour and capital as the main factors of production missed one of the most important agents of economic change – innovation. It is the capacity to innovate that has started restructuring the world's industry. The comparative advantage theory led to the view that the availability of cheap labour in the Third World countries would ensure their participation in global industrial production. These countries formulated policies aimed at attracting foreign investment; profit repatriation was guaranteed, wage increases discouraged, union activities suppressed and low tariffs on imported machinery instituted.

The analogue of this view prevailed in the agricultural sector. It was believed and advocated that the natural endowment of the tropical countries provided them with comparative advantage over their temperate counterparts in the production of certain agricultural commodities. The theory, of course, considered innovation to be exogenous to the production system, as supposedly intrinsic factors such as climate and soil fertility played the dominant roles. Combined with the availability of cheap agricultural labour, the comparative advantage theory held firm in the minds of agricultural planners and policy analysts. Even when the terms of trade were sliding against agricultural commodities, international debates focused on introducing schemes that would stabilize prices – a truly Newtonian programme to establish some form of 'balance' between the industrialized and raw material producing countries.

But major biotechnology firms have turned their attention to products which irreversibly substitute for imports from Third World countries. The objectives of these firms are summarized by the case of a leading US biotechnology firm, Calgene:

> The value of most crops grown for food processing is determined primarily by the crops' processing characteristics such as texture, flavor, color, protein and carbohydrate content, and shelf life. Food processors have traditionally bought raw materials in commodity markets where all products are essentially undifferentiated. With recombinant DNA technology, however, Calgene can provide food processors an opportunity to gain competitive advantage by allowing precise genetic modification to develop proprietary crop varieties with enhanced characteristics which can then be patented and grown for their exclusive use.[35]

Calgene, like other biotechnology firms, has realized that technological advancement is the main source of competitive advantage. The focus is to use biotechnology to substitute for some of the key imports of the industrialized countries. Nearly all the major high-value crops exported to the industrialized countries have become targets for displacement through biotechnology research. The attention has been directed at high-value plant products. An increasing number of natural products will, in the near future, be produced through techniques such as rDNA, cell fusion and tissue or cell culture. This partly explains why large chemicals firms are building their in-house capability in biotechnology while at the same time supporting a large number of research projects on contract basis.

In order to understand the potential impacts of these changes, it is necessary to adopt an alternative analytical approach. A non-equilibrium approach would show that it is the destabilizing introduction of new knowledge and technology that reorganizes production. The application of biotechnology to agricultural production illustrates this point. Biotechnology may have two effects on agricultural production. In the first place, it may shift production to other Third World countries which have already taken the lead in the use of biotechnology. Secondly it may also enable Third World countries to deal with the effects of agricultural relocation, an option that will be discussed later (see Chapter 7). The case of cocoa shows how the application of biotechnology is likely to relocate production from Africa to Malaysia and Brazil. Other examples from the sweetener, flavours and insecticide industries show how production is being relocated from Third World farms to industrialized country laboratories. From these cases it is obvious that what is important in long-term economic production is not simply factor endowment but the capacity to innovate; those who have competitive advantage in technological know-how are the ones who will define the future of global agriculture.

International cocoa trade owes its origin to the invention of milk chocolate in 1876.[36] Over the last century cocoa has become the second most important tropical commodity in international trade. Like other raw materials, cocoa is produced in Third World countries and processed in industrialized countries. Most of the world's cocoa is produced in the West African states of Côte d'Ivoire, Ghana, Nigeria and Cameroon, which represent 55 per cent of the world market. Over the 1980-85 period, the seven leading cocoa producers in the world (Côte d'Ivoire, Ghana, Nigeria, Cameroon, Brazil, Ecuador and Malaysia) accounted for less than 5.0 per cent of the world's export of chocolate and chocolate products, while six consuming countries accounted for 53 per cent of the export.[37]

The global distribution of cocoa is dependent on key climatic and geographic factors which may suggest that only a few countries are endowed with the natural potential to support the plant. Cocoa tends to thrive well in areas which receive between 1,150 mm and 2,500 mm of rain a year. Excess rainfall may result in disease, particularly black pod (phytophthora) and cocoa will not grow well under deficient rainfall. The nature of the soil affects the quality of the cocoa and the plant cannot endure temperatures less than 10°C and more than 32°C. The ideal range is between 21°C and 32°C. Finally, windy areas are unsuitable for growing cocoa.

In addition to these climatic limitations, the cultivated cocoa is based on a very narrow genetic range which makes the plant vulnerable to disease and pests.[38] It is estimated that some 20-30 per cent of the world cocoa output is lost to pests and disease.[39] Phytophthora pod alone is estimated to destroy 10 per cent of the world cocoa output. Substantial losses are also caused by witches' broom and canker and swollen shoot virus. Research on these diseases is conducted by the producers and consumers alike. Some of the research has resulted in the introduction of hybrid cocoa trees which yield 1.0 tonnes per hectare per year instead of the traditional trees which produce 0.3-0.5 tonnes per hectare per year.

The introduction of hybrids has resulted in the reduction of land devoted to cocoa in Nigeria and Ghana, since the world consumption of cocoa beans has risen by a modest 1.8 per cent over the last 20 years.

Table 4.9
Research in high-value natural substitutes

Plant	Product	Origin	Research Organization	Value (US$/kg)	World Market (US$ million)
Acacia	Gum arabic	Nigeria, Senegal, Sudan	TIC Gums (USA)	—	60
Catharanthus	Vincristine	Mozambique, Israel, India, Sri Lanka, Thailand, Vietnam, Madagascar	Canadian National Research Council Eli Lilly	5,000	—
Cinchona	Quinine	Indonesia, South America, Kenya, Tanzania	Plant Science (UK)	—	—
Cocoa	Cacao butter	Ivory Coast, Ghana, Brazil, Nigeria, Malaysia, Equador, Cameroon	Cornell University, Hershey (US), Nestlé (Swiss), DNA Plant Technology (USA), Genencor (USA), Ajinomoto (Japan)	8.8	2,000
Digitalis	Digitoxin Digoxin Ditoxin	Eastern Europe, India	University of Tübingen (FRG) Boehringer-Manheim (FRG), Plant Science (UK)	3,000	—
Dioscorea	Diosgenin	Mexico, India, China	—	—	—
Jasminum	Jasmine	—	—	500	—
Lithospermum	Shikonin	Korea, China	Mitsui Petrochemicals (Japan)	4,500	—
Papaver	Codeine Opium	Turkey, Thailand	Plant Science (UK)	650	—
Pyrethrum	Pyrethrins	Tanzania, India, Kenya, Equador	University of Minnesota (US) Minnesota (US)	300	110
Rauwolfia	Reserpine Rescinnamine	India, Mozambique, Thailand, Zaire	—	400	80
Sapota	Chicle	Central America	Lotte (Japan)	—	—
Thaumatococcus	Thaumatin	Liberia, Ghana, Malaysia	Tate and Lyle (UK), Unilever (UK/Holland), INGENE (US), DNA Plant Technology (US)	2,200	—
Vanilla	Vanilla	Madagascar, Indonesia, Comoro Is., Reunion	David Michael (US), International Plant Technology Institute (US), DNA Plant Technology (US)	70	66

Sources: Kenney and Buttel, 'Biotechnology'; Rural Advancement Fund International, Pittsboro, North Carolina, USA; International Trade Centre UNCTAD/GATT, Geneva.

Nearly half the world's cocoa is grown by smallholders, but the international trading is carried out mainly by state firms. A combination of numerous smallholders and government involvement in cocoa trade makes it difficult for the importing countries to have effective control over the supply. Moreover, the narrow agro-ecological region in which it can grow makes the chocolate industry dependent on a few suppliers who, in turn, can easily influence the market. The political nature of the cocoa trade is exemplified by recent efforts to ratify international cocoa agreements. Three attempts have been made since 1972. The most ambitious was the 1980 International Cocoa Agreement which failed partly because it was neither ratified by the largest producer, Côte d'Ivoire, nor the largest consumer, the US. Subsequent negotiations led to another agreement in 1986, which was ratified by Côte d'Ivoire and is on provisional operation.

The growing politicization of the international cocoa market, and economic uncertainty in the producing countries, have led to the need for the importers to diversify their sources. The need could, however, be met only with new ways of producing cocoa in areas and from varieties that were not possible in the past. Biotechnology has provided new possibilities for producing cocoa varieties that yield high quality cacao butter, higher yielding trees, and greater pest and disease resisting plants. New approaches to convert, through enzymatic processing or protein engineering, cheap soybean or palm oil into cacao butter and to produce cacao butter through cell culture are also being investigated.

In 1986 the Pennsylvania State University (Penn State) initiated a programme to study the molecular biology of *Theobroma cacao* with a US$1.5-million endowment from the Chocolate Manufacturers Association of the United States and its research wing, the American Cocoa Research Institute. The programme is aimed at using biotechnology to stabilize the export crop for the chocolate producers. Penn State has started creating a 'cacao gene library' from which material for genetic engineering will be obtained. The researchers will endeavour to produce high-yielding and high-quality cocoa plants through the introduction of new varieties which have 'more cocoa pods on each tree, more beans in each pod, larger beans of uniform quality, and trees resistant to drought, cold, fungi, viruses and pesticides.'[40]

One of the challenges in cocoa breeding is micropropagation, or the regeneration of large numbers of genetically identical cocoa plants in the laboratory. This process is much faster than the conventional methods used by plant breeders. After this is achieved, it will be possible to replace large sections of the cocoa farms with new varieties in a short period. Another item on the Penn State research agenda is to increase the fat content of cocoa beans so as to yield more cacao butter. With the help of genetic engineering, the scientists hope to incorporate the sweetness (thaumatin) gene of the West African plant, katemfe (*Thaumatococcus daniellii*) into cocoa plants, which would eliminate the need for sugar.

Thaumatin is a protein and therefore the resulting chocolate would be sugarless.

In addition to university-based research, small biotechnology firms are also working on new ways of diversifying the source of cacao butter. DNA Plant Technology of Cinnaminson, New Jersey (USA) is already working jointly with Hershey Foods, the largest US chocolate manufacturer, to produce new varieties of cocoa using cellular genetics and tissue culture techniques.[41] Work on cocoa is also being undertaken by Cadbury-Schweppes at the Lord Zuckerman Research Centre in association with Reading University, UK.

While these research activities are underway, other firms are working on ways of converting other vegetable oils to cacao butter. A Tokyo University researcher has patented a process to produce cacao butter substitutes by enzymatically processing cheap oils. The process has already been licensed by a major Japanese food company, Ajinomoto of Tokyo. A similar process has been developed by Genencor of California. Genencor intends to put fat-producing enzymes on the market by the early 1990s. The company is owned by Genentech (California), Corning Glass Works (Corning, New York), A. E. Staley (Illinois) and Kodak (Rochester, New York). In another approach, CPC International (New Jersey) has a patent on a process that produces a cacao butter-like substance from growing yeasts with fatty acids.

The production of cacao butter using tissue culture techniques has been a major item on the corporate research agenda. This option is attractive because it replaces the land requirements for the production of cacao butter, since the process would be undertaken in a laboratory. So far attempts by Cornell University with the support of Hershey Foods have not been successful, partly because the process does not seem to yield the right fatty acids. The cost of producing cacao butter in the laboratory is about US$220 per kilogramme, compared to US$4.0 per kilogramme of cocoa beans. The incentive to produce cacao butter through tissue culture will depend on whether alternative methods will be available. At the moment, efforts to produce high-yielding varieties will further reduce the price of cocoa beans thereby undercutting the incentives to use tissue culture techniques.

It is difficult to predict the exact effects of these innovations on the future cocoa markets. It is possible, however, to identify some of the trends that might emerge as a result of the availabilty of high-yielding varieties. The expected yields of 1.36 tonnes per acre can be achieved only through intensive cultivation methods. Malaysia has already reported yields of up to 2.0 tonnes per hectare from hybrids. Countries such as Malaysia which are already producing hybrid cocoa and using labour-reducing methods are more likely to adopt these varieties faster than their African counterparts. Brazil has an active programme in biotechnology and may adopt the new methods of producing cocoa varieties earlier than the African countries. In addition, the production may shift from African countries dominated by smallholders to countries such as Malaysia and Brazil which have large farms.

The impact of these changes on the African economies is likely to be profound. First, the changes will entail the loss of substantial foreign exchange earnings. Second, the relocation of agricultural production will undermine employment opportunities in the smallholder agricultural sector. Third, the use of the new varieties is likely to be controlled through patent or variety protection. Even if the African countries adopt the new varieties and use them in smallholder production, the possibility for future tissue culture innovations to undermine farm production is still a major threat. It is interesting to note that despite these possible adjustments in the world cocoa industry, the African producers have made no definite policy decisions on how to deal with the changes and some of them, such as Uganda, are planning to revive or expand cocoa production.

Even sections of the UN responsible for providing trading guides on Third World commodity markets such as the International Trade Centre (ITC) which is operated jointly by the UN Conference on Trade and Development (UNCTAD) and the General Agreement on Tariffs and Trade (GATT) have so far failed to provide reliable analyses of the potential impact of biotechnology on the world commodity market. For example, ITC's 1987 report on cocoa does not review the possible impact of biotechnology on the cocoa industry although its main objective is to 'provide developing countries with up-to-date information on the world cocoa market'.[42] This may be an oversight on the part of ITC but it also reflects the fact that such established institutions still apply conventional analyses that treat innovations as exogenous to the economic process. The study, for example, relies largely on statistics on supply and demand and does not adequately account for the impact of innovations on the cocoa market. While such studies may be useful for short-term trading arrangements, they are of little value in guiding Africa over the long run, especially in relation to the major restructuring of agricultural production likely to result from innovations in biotechnology.

The impact of biotechnology research is not only going to affect foreign exchange earnings from raw material exports, but it is also going to erode the agro-industrial prospects of these countries. Those countries which have already invested in downstream processing are likely to experience the premature scrapping of the machinery and the accumulated knowledge as well as skills; they are facing major discontinuities in their efforts to accumulate industrial capability.

These trends are further exemplified by the case of vanilla. The plant, *Vanilla planifolia,* is indigenous to Central and South America and it is now largely grown in Indonesia, Madagascar, Comoro Islands and Reunion. These countries earn a total of US$67 million annually from vanilla exports. The last three of them account for some 98 per cent of world vanilla output. Madagascar alone meets about 75 per cent of the world's vanilla market, a production process that involves over 70,000 smallholders. The country's annual earnings from the crop are US$52 million (about US$6.0 million goes to Comoro Islands and US$1.0 million to Reunion). Madagascar is dependent on vanilla for up to 10 per cent of its annual export earnings. Some 60 per cent of Comoros' foreign exchange earnings are accounted for by the export of vanilla. The potential loss of Comoro's vanilla export market is proportional to the loss of Kenya's export markets for coffee and tea. The US is the main importer of vanilla, accounting for 58 per cent of the world's market. Its 1985 import bill for vanilla was US$47 million.[43]

Current biotechnology research is destined to undercut these export earnings by producing vanilla flavours through tissue culture techniques. Present prices for using tissue culture to produce vanilla are still expensive and the product costs about US$2,200 per kilogram while the current price for vanilla beans on the market is about US$70.5 per kilogram. New techniques being developed will reduce the price of vanilla produced by tissue culture to less than US$50 per kilogram, thereby undercutting agricultural producers. One of the attractions of tissue culture production is that the process would be controlled industrially and the market would be freed from political and social sources of price fluctuation. Vanilla research is currently being undertaken by the US firm, David Michael of Philadelphia with collaboration from researchers at the University of Delaware. Similar work is also underway at the International Plant Research Institute (IPRI) at San Carlos, California. The firm intends to put a product on the market in 1989.

A wide range of natural sweeteners are now being discovered and considered for large-scale production. The 'sweetness gene' of katemfe has been inserted into bacteria, making it possible to produce natural sweeteners in vats as an industrial product. The plant contains two sweet-tasting proteins, thaumatin I and II, each of which is 1,600 times sweeter than sucrose (on weight basis).[44] On a molecular basis, thaumatin is 100,000 sweeter than sucrose. Thaumatin is marketed by Tate and Lyle in Britain, Austria, Switzerland and Japan under the trade name of Talin and used in candies, chewing gum, yogurt, coffee, soft drinks, animal feed, pet food, pharmaceutical products, tobacco products, pickles, jellies, soups and other products. It costs US$2,200 per kilogram.[45]

Under the current techniques, thaumatin has to be extracted from plants growing in their natural environment. Tate and Lyle established katemfe plantations in Liberia, Ghana and Malaysia in the 1970s while carrying out research to relocate production into laboratory vats. But by using genetic engineering techniques, firms can now produce thaumatin industrially. Firms working on this approach include International Genetic Engineering (INGENE) California, which developed a genetically-modified thaumatin protein in 1982 and plans to apply for approval from the US government to market the product in 1988. INGENE has been working under a contract for Beatrice Foods, Chicago, and holds a patent on regulatory genetic sequences for producing thaumatin.

DNA Plant Technology, which is working on vanilla, has also announced a research programme to produce thaumatin using cell culture technology. The firm is also working on other major exports of Third World countries such as palm oil, tomatoes, coffee, cocoa, fragrances and flavours. Unilever, which was the first firm to express genes for the thaumatin protein in microbial hosts, has continued to work on the sweetener.

It is significant to note that work on thaumatin has attracted some of the largest firms in the world which are working under contract with universities or emerging biotechnology firms such as DNA Plant Technology and INGENE. Beatrice Foods is a major food corporation whose annual sales for 1985 amounted to US$12.6 billion. It was ranked 26 in *Fortune*'s 500 largest industrial firms in the world. Unilever's 1985 sales amounted to US$21 billion and ranked 18 in the *Fortune* 500. INGENE and DNA Plant Technology were founded in 1981 and have concentrated on applying genetic engineering and tissue culture techniques to the manufacture of pharmaceutical, food and speciality chemical products.

Thaumatin is one of the numerous substitutes for sugar that have begun to change the world commodity market irreversibly; there are other candidates. The miraculous berry (*Synsepalum dulcificum*) from West Africa is a protein-based candidate that might lead to a new range of non-fattening sweeteners; this berry contains negligible calories. Also from West Africa is the serendipity berry (*Dioscoreophyllum cumminsii*) which contains a protein called monellin which is 3,000 times sweeter than sucrose. From Paraguay is bertoni (*Stevia rebaudiana*) whose leaves are 300 times sweeter than sucrose; it is already being marketed in Japan. A Chinese fruit, *Momordica grosvenori,* sold in southern China and Hong Kong, is a potential candidate for large-scale sweetener production.

The new hunt for natural sweeteners has also led to the rediscovery of *Lippia dulcis,* which the indigenous South Americans reportedly enjoyed chewing at the time of the conquest of the Aztecs; this candidate is 1,000 times sweeter than sucrose. So far the commercialization of natural sweeteners has been limited to a few products, and their impact is yet to be felt on a large scale.

There are several factors which indicate that the sweeteners will pose long-term challenges to the traditional sugar industry. The natural sweeteners do not have uniform taste so they will offer diversity in the food industry, whose growth is largely influenced by product and taste differentiation. The use of genetic engineering makes it possible to extract the relevant genes from the plant and insert them in another organism, thereby freeing production from the complicated agricultural process associated with political and market uncertainty. This will directly affect farm employment in the sugar-cane sectors of the Third World countries. It is estimated that up to 50 million people in the Third World countries depend on world sugar markets for their livelihood.

Countries such as Kenya have in recent years devoted some of their sugar-cane production capacity to meeting local market demand. The local market, however, is not free from competition from biotechnology-based sweeteners, especially for industrial use. Numerous soft drinks firms, including Coca Cola, have already switched from sugar to substitutes for some of their products in the US and it is a matter of time before these products reach the African countries. It may require deliberate government policy to protect the local sugar-cane production capacity from such competition. The decision will have to take into consideration the need to save the current jobs provided by the industry. Given the very high levels of sweeteness, even extremely low conversion efficiencies would yield substantive amounts of sweetener. The discovery of protein-based sweeteners is likely to create a new market among those concerned with the health effects of sucrose or saccharin. And of course the search for new varieties opens up possibilities for the discovery of more suitable substitutes.

The direction of this research is aimed at reducing dependence on imported raw materials. The impact on the countries exporting this product will be profound and irreversible. Over the years, large sections of the population have organized their lifestyles around the production of these crops. This is going to be changed by current developments in biotechnology. The impacts will not be only economic, but will have long-run political implications and the attempt by communities to reorganize themselves in response to the changed conditions. It is, therefore, in the interest of raw material exporters to closely monitor current trends in biotechnology and the use of genetic resources and modify their internal policies in anticipation of potential long-term effects.

Another example is that of water soluble gum arabic which is extracted from *Acacia senegal,* a tropical tree. Sudan accounts for 80 per cent of the world's output of gum arabic which is used in the beverage and confections industry. In 1983, the product accounted for about 8.0 per cent of its foreign exchange earnings, a total of US$57 million. Nearly 30 per cent of the world's gum arabic output is consumed by the US, which imported 5,140 tonnes valued at US$18 million in 1984. This export is currently threatened by substitutes, most of which are blends of gum arabic and starch compounds. The substitutes were partly a result of supply shortages over the 1984/85 period. Since substitutes emerged when there were supply shortages, the producer countries were not able to assess the trends on the demand side. Moreover, the high prices associated with the short-term supply shortages stimulated interest in increasing production precisely at the time when it was not advisable to do so due to the advent of substitutes.

Once substitutes are introduced, it is difficult to withdraw them. It is important to recognize that the introduction of locally-produced substitutes for imported raw materials is always associated with political support, and institutional as well as legal reforms which make the changes irreversible. Moreover, such developments are related to the need to maintain self-sufficiency in speciality products and are therefore supported through a wide range of government policies, which include subsidies, legal reform, tariff and non-tarrif barriers, government procurement, R&D funding, tax rebates and other financial facilities.

The case of pyrethrum further illustrates the restructuring of the global agricultural sector. The history of pyrethrum should serve as a local example of the impact of technological developments on raw materials. Kenya is one of the main exporters of pyrethrum flowers and extracts and has been affected by the changes in the international market as well as technological advances. Kenya's export earnings from pyrethrum extracts declined from K£3.0 million (at current prices) in 1969 to K£1.3 million in 1984. Over this period the share of flowers declined to 50 tonnes as the country switched to extracts. Production remained sluggish in the mid-1980s. The 1986 production saw dramatic improvements as world prices improved. The increase was also associated with a crop improvement programme.

Kenya has been exporting a large share of its pyrethrum to the US, which now accounts for 70 per cent of Kenya's export market. This market is destined to be undercut irreversibly because of biotechnology research, especially through the use of tissue culture to produce pyrethrins. Most of the major research has been conducted by the University of Minnesota with funds from McLaughlin Gormley King, a major US importer of pyrethrum. The current US pyrethrins market is valued at over US$20 million and a kilogram fetches nearly US$300. This amount is large enough to attract biotechnology investment. The entry of biotechnology-based pyrethrins will irreversibly reduce the market share currently enjoyed by exporters in Kenya, Tanzania and Ecuador. Ironically, the genetic material on which the research is based was supplied by the countries involved in the production of pyrethrum.[46] It is not certain that these countries were aware of the long-term impacts of technological development in pyrethrum tissue culture research.

The restructuring of global agriculture through the use of biotechnology is starting to be obvious at a very early stage of the growth of the industry. Given the pace of technological change, the impact of the innovations of the early 21st century will be even more profound. The underlying theme here is that it is those who have a competitive advantage in scientific and technological knowledge who will shape the global agricultural sector. Land and labour are becoming less relevant than they were when the current economic theories were conceived. Those who control technology also control the global flow of investment capital.

The shift towards science-intensive agricultural production is changing the international distribution of sources of raw material as well as the patterns of international trade. This reorganization of the global agriculture will also change both the locus and process of international negotiations. The availability of a wide range of biotechnology-based plant varieties will make it possible for the industrialized country producers to maintain extensive flexibility in the quality, quantity and prices of raw materials. In addition, these countries will reduce their dependence on a handful of major producers. This will weaken the possibility for collective bargaining or action among the producers. International commodity agreements are likely to be less effective than they are normally expected to be.

Traditional negotiation forums such as the UN are also likely to become less effective and therefore countries might tend to enter into bilateral agreements on certain commodities. This will create conditions that will make the African countries more vulnerable to trading pressures from the industrialized countries, especially given their limited negotiating capability and regional collaboration. Institutions such as the GATT are likely to become more effective in imposing the will of the industrialized countries on unorganized commodity producers. These changes may, in turn, force the African countries to look into new forms of regional collaboration. At the moment it is difficult to say which kinds of institutions are likely to emerge.

Advances of conventional plant breeding methods in conjunction with advances in biochemistry helped to mould the current shape of world agriculture. This shape is on the verge of being redesigned with the help of new developments in genetic engineering, tissue culture and cell fusion techniques. There are numerous opportunities for producing new plants and animals which will introduce significant discontinuities and convergences in the industrial and agricultural sectors.

The introduction of new life forms into the economic sphere is likely to lead to major economic changes. These changes, however, are not likely to occur without major change in the institutional organization of production at national and international levels. One of the ways of ensuring that the emerging technologies will be introduced without undue obstacles is to reform the existing legislation to respond to the imperatives of technological change. Legal reform has often co-evolved with the technological advances. This trend has been salient in relation to changes in the field of biotechnology. It is notable that most of the major battles in this field have been fought on the legal front. The legal regime has been gradually giving biotechnology more room for adaptation.

Notes

1. See Faulkner, 1986, for details.
2. See Juma, 1986, for a history of fermentation technology.
3. Penicillin and numerous enzymes are products of this generation.
4. Significantly the logic behind patent systems resembles the Spencerian view of progress. The system is viewed largely as a process of selecting good inventions from unsuitable ones on the basis of industrial applicability, novelty and inventiveness. Patents that are not worked over a certain period are revoked. This suggests that the usefulness of an invention is deterimined over the short period following its registration. In the final analysis, it is the market that is left to determine the usefulness of any invention. On the contrary, a systems view would show that major technological recombinations may occur between inventions that may have been deemed valueless. There is also a tendency to use the Spencerian view of progress, which assumes that the shift is always towards higher forms of technical change. The view of invention as a solution to technical problems is too instrumental because it does not consider the process as part of social learning. There are major discoveries that are not solutions to any obvious or prevailing technical problems; for a long time fibre optics was known as a solution looking for a problem.
5. Murray, 1986, p. 293.
6. Ibid.
7. Doyle, 1985, pp. 25-6.
8. RAFI, 1987e, p. 6. It should be noted that one of the features of the Green Revolution was the ability of the breeding programmes to provide varieties that were responsive to fertilizers and therefore increased the profit margins of agrochemical firms.
9. RAFI, ibid., pp. 2-3.
10. Such a research programme also allows for the development of herbicides that do not adversely affect the environment. But whether such criteria will be taken seriously depends largely on the costs involved, existing legislation and market factors.
11. 'Resistance traits usually take one of four forms. In one form, the plant does not absorb the herbicide. In another, the plant does not transport the herbicide into those tissues where it can do damage. Or the plant may detoxify

the herbicide by breaking it down. Finally, the site within the plant's cells where the herbicide would act may not be susceptible', Benbrook and Moses, 1986, p. 56.

12. 'While such toxins have the practical ecological advantage of being specific for certain insect species, they now have to be applied to crops in costly spraying programs. The production of such proteins within the cells of genetically engineered plants might provide pest resistance at both reduced cost and with improved environmental safety over present control measures', Barton and Brill, 1983, p. 674.

13. Advanced Genetics Sciences is seeking to commercialize the technology, which was developed from research conducted at the University of California, Berkeley.

14. OTA, 1986c, pp. 34-5.

15. Zedan, 1987. The Kenyan MICERN has developed a *Rhizobium* inoculant that could reduce considerably the used chemical fertilizers if used widely. Other experiments using *Frankia* have been tried in Tanzania, Egypt, Senegal, Zaire, Rwanda, Somalia, Gambia, Nepal, Sri Lanka, Indonesia, China and Costa Rica.

16. For details on the functioning of the MIRCENs, see Zedan and Olembo, 1985.

17. One research project being undertaken under the auspices of the programme has resulted in a reduction of soy fermentation period from three months to nine days while reducing the salt content by 50 per cent.

18. Kenney, 1987, p. 12.

19. As part of its increasing work in biotechnology, China has launched a journal entitled *Chinese Journal of Biotechnology.*

20. Juma, 1985, p. 47.

21. Blumenthal et al., 1986, p. 1364. Despite the findings of the survey, Blumenthal et al., argue: 'Any losses to science or to university values that result from marginal increases in the level of secrecy in universities may be more than offset by net additions to knowledge that result from the infusion of industry funds into the labs of talented faculty. Marginal shifts in the direction of university work toward more applied and commercially relevant projects may have benefits for human health and economic growth that far outweigh the risks of scientific progress', ibid., p. 1365. Contrary evidence is given in Nelkin, 1982.

22. Kenney, 1987, p. 5.

23. See Bohm and Peat, 1987, pp. 63-103.

24. OTA, 1987a, p. 90.

25. Fogleman, 1987, p. 189.

26. Sharples, 1987b, p. 96. See also Sharples, 1987a, and Odum, 1985. A conflicting position is provided in Davis, 1987.

27. For details on weeds, see Holm et al., 1987.

28. Colwell et al., 1985, p. 111. See also Levin and Harwell, 1986.

29. See OECD, 1986, for some of the effects of biotechnology. The concern over the risks of recombinant DNA, especially for researchers, is illustrated in Bartels, 1986.

30. Joyce, 1986, p. 15.

31. For a review of Japan's biotechnology programme, see Saxonhouse, 1985.

32. Webb, 1987, p. 222.

33. 'Time is of the essence in taking advantage of European financing opportunities. The first wave of U.S. companies is likely to get the best deals, as EEC governments strive to prime their biotechnology parks and catalyze long-term strategic investment in new manufacturing facilities', ibid. p. 229.

34. Regal, 1985, pp. 4-5.

35. Calgene, 1985, p. 11. See Balandrin et al., 1985, for a survey of technical possibilities for natural substitutes.

36. Cocoa (*Theobroma cacao*) is native of the Amazon basin and was already being used by the Aztecs as a medium of exchange and a beverage when the Spanish arrived in South America.

37. ITC, 1987, p. 23. These countries produced 84 per cent of the world cocoa in 1985.

38. Numerous expeditions to collect cocoa plants were launched in the 1980s, mainly with the support of the cocoa industry. Some of the expeditions were undertaken through IBPGR.

39. ITC, 1987, p. 11. Some estimates put the loss at 50 per cent.

40. Shand, 1987, p. 3.

41. Ibid., p. 4.

42. ITC, 1987, p. 1.

43. RAFI, 1987a.

44. Myers, 1983, p. 166.

45. Tate and Lyle market the product in formulations of 1.0 per cent and 10 per cent Talin concentration in liquid or powder form. For a review of recent developments in alternative sweeteners, especially on high fructose corn syrup (HFCS) see Thomas, C. 1985.

46. Zieg et al., 1983, pp. 88-91.

5. Life as Intellectual Property

The private ownership of genetic resources has been a major source of international controversy. Not only have improved plant and animal varieties been subject to patenting and patent-like protection, but their genes are also now being patented. The US was the first country to grant patents for plants. Other countries maintained a demarcation between utility patents and plant breeders' rights (PBRs). This demarcation was based on a false dichotomy between inanimate and biological innovations. The early plant introduction activities in the US were undertaken by the Patent Office. Today, the patenting of life forms shows that the rise of PBRs was only a temporary institutional development which was not based on a clear understanding of the process of technological innovation. The close linkages between technological change and institutional innovation are clearly reflected in the recent reforms in intellectual property law relating to biotechnology.

Patents and innovation

A patent is a legal right which confers exclusive rights for an invention over a limited period. The patent system has evolved over a long time from similar forms of granting monopoly rights to traders or inventors. Greek records show that monopoly rights were granted as early as 200 BC. The Romans granted monopolies to inventors and merchants in the form of 'letters patent', derived from the medieval Latin *litterae patentes,* or open letter addressed to the public.[1] But these rights were later abused as kings granted monopolies to favoured people who kept the prices of the commodities higher than they would normally have been. Subsequent British modifications in the system stopped such monopolies, except on inventions.

Patent rights are granted in exchange for the disclosure of the invention. This disclosure is expected to serve at least three functions. In the first place, it helps to prevent other people from infringing the patent during the term of monopoly. Second, it allows the public to have access to the patentee's secrets on the expiry of the patent duration. Third, it helps to prevent patents from being granted for already known inventions. Most countries require the disclosure to be done in such a way that the invention can be reproduced or worked by a person who is skilled in the trade. This is a significant requirement because it is from such disclosure that the society is expected to gain from the availability of the invention.

Most laws require new inventions to fulfil the requirements for novelty, inventive step and industrial applicability. An invention is considered novel if it does not constitute the state-of-the-art and in order to establish this it is necessary to undertake a novelty search, mainly through a patent office. The availability of information on the state-of-the-art may save large amounts of money that would otherwise be invested in research on technical solutions that already constitute the state of the art. The Swedish Patent Office, for example, rejects about 2,000 patent applications a year for lack of novelty, the equivalent of US$50 million spent on research.[2] If an invention is not obvious to a person skilled in the art, it is said to constitute an inventive step. The requirement for industrial applicability is aimed at ensuring that society does not engage in 'useless' inventions but those that contribute to economic growth.

Economic development in Europe co-evolved with significant technological innovations that led to the consolidation of intellectual property rights as a fundamental social ethos. Part of this was reflected in the growth of intellectual property as a basic right, especially in inanimate inventions. Let us examine the development of intellectual property in West Germany as an example. The German patents system originated in the 11th and 12th centuries when production and marketing were organized under the craftsmen guild system. The guilds provided monopoly for craftsmen products and later became examination boards for determining whether inventions were new or useful.

The first regional patent-like privileges were granted in 1378 by feudal dukes in the mining industry for pumping water from sub-surface mines and constructing tunnels in the mines. Up to the 17th century, the emerging German patent system separated the remuneration of the inventor from monopoly rights granted to inventors. The absolutist political system that prevailed in Germany in the late 17th and early 18th centuries combined the compensation for invention and the use functions of the invention. This marked the beginning of the active stimulation of industrial production by the state.

The modern German patent law was passed in 1877 and defined patentable improvements as new inventions which permit commercial use. Food, pharmaceutical and similar products were excluded from patenting. Exclusive use was also restricted where the army, navy or any other public interest required the use of particular inventions. Patents could be revoked if, among other reasons, they were not worked in three years. The law required that the patent information disclosed should be sufficient for technically skilled people to be able to produce the said innovation.

European and American advocates of the patent system participated in the International Convention for the Protection of Industrial Property held in Paris in 1883. The Paris Convention marked a major step in the internationalization of intellectual property rights. At this period, the notion of national novelty was being replaced by the principle of international novelty. The patent system, like other institutions, had to respond to the imperatives of international trade. Following the growth of the international division of labour and expansion in trade, Germany found it necessary to seek international protection and recognition of its inventions. Efforts were therefore undertaken to harmonize the German patent laws and introduce new ones that would promote German interests in the international market.

Patents are justified under the German system by four main arguments. First is the 'natural law', which states that inventions should be the inalienable property of the person who came up with the idea, and who should be compensated. The second is the idea that the inventor needs to be remunerated for his contribution to society. The third is a contract between the public and the inventor, where the first grants temporary monopoly while the other grants disclosure of the invention. The fourth is the argument that patents induce inventive activity.

The role of patents in industrial development has been a subject of extensive study. It is difficult to establish a direct relationship between economic growth and patent activity, although patents are used as an indicator of innovation and economic development. Countries such as Japan have been able to make major advances in certain fields despite a relatively low rate of patent activity in those fields. What is important though, is the capacity of a country to introduce inventions into the economic system. This issue is more a matter of the nature of the prevailing technology policy and does not necessarily relate to patents in a causal way.

The late 19th century marked a critical point in the consolidation of the European powers in the international economic system. The Berlin conference, in 1884-85, legitimized the imperialist claim of various European powers to particular territories. It is not a coincidence that the Paris Convention was signed in the same period. A wide range of developments, such as colonization, the internationalization of trade in manufactured goods and the expansion of colonial agriculture, were part of a wider expansionary process among the Western countries. By then, most of the major plants that were later to dominate colonial agriculture had been identified and even established in some colonies. By the turn of the century, the stage was set for the subsequent introduction or expanded production of commercial plants in African economies: rubber in the Congo; ground-peanuts in West Africa; tea in Malawi; sisal in Tanzania and tea as well as coffee in Kenya, to name just a few.

The rise of plant breeding and the contribution of agriculture to economic development led breeders to seek property ownership over the products of their effort. The need to protect plant varieties, as we shall see later, was based on the view that breeders invested time and resources into the process and needed to be rewarded. It was also felt that breeders were like inventors and their contribution to society needed to be recognized. Indeed, the early stages of industrialization were also marked by increased agricultural production partly to feed the growing urban population.

Plant breeders' rights

Plant breeders' rights (PBRs) are property rights granted to breeders to afford them exclusive rights over new plant varieties. There are several reasons why PBRs are granted. Plant breeders have over a long period felt that the products of their labour should be treated in the same way as inanimate inventions. This means that the reasons for granting inventions as outlined above are also applicable to new plant varieties. In the first place, breeders who produce new varieties that have market potential feel that they should be compensated for their contribution to society in the form of a royalty. Second, the granting of PBRs, like that of patents, is expected to stimulate further plant breeding.

In essence, PBRs provide protection against farmers, public sector breeders and other private breeders. They grant exclusive rights over a given period in which the breeders can be protected against competition and the unauthorized use of the protected varieties. Such protection is felt necessary partly because breeders invest a large sum of money in breeding activities and therefore need some guarantee from the government that the costs will be recovered. In addition, it takes a long time to produce new varieties and put them in production. The process normally goes through several stages and the average time spent on new varieties depends on crop types and regulatory factors.

The history of PBRs represents an interesting case in the co-evolutionary development of international institutions. The seemingly divergent development of patents and PBRs shows that the two systems were separated, partly as a result of the limited understanding of the process of technological innovation. It should be noted that in the early days of the evolution of the US agricultural system, plant introduction was dealt with by the Patent Office; germplasm collection was done in conjunction with the gathering of agricultural information and technology.

Table 5.1
Time required to develop new varieties

	Years			Number of Varieties*	
Crop	**Cross to Date of Determination**	**Date of Determination to Application**	**Total**	**Cross to Date of Determination**	**Date of Determination to Application**
Barley	7.0	3.4	10.4	9.0	10.0
Beans	8.0	3.3	11.3	16.0	32.0
Cauliflower	11.0	7.5	18.5	1.0	4.0
Corn	5.5	7.5	13.0	4.0	6.0
Cotton	8.0	4.2	12.2	27.0	57.0
Oats	8.8	2.1	10.9	13.0	16.0
Onion	9.0	2.9	11.9	11.0	14.0
Peas	7.0	4.0	11.0	24.0	34.0
Rice	6.0	2.8	8.8	5.0	12.0
Safflower	6.0	1.7	7.7	4.0	5.0
Soybeans	6.2	3.0	9.2	64.0	75.0
Squash	11.0	3.7	14.7	1.0	3.0
Tobacco	8.5	2.6	11.1	11.0	14.0
Tomato	8.3	1.4	9.7	3.0	9.0
Watermelon	8.5	5.0	13.5	2.0	10.0
Wheat	8.0	2.8	10.8	36.0	56.0
Average/Total	7.9	3.3	11.1	253.0	391.0

* Applicants are required to list the date of variety determination and date of application when submitting protection applications. They are not required to list the date the cross was made. It is for this reason there are fewer varieties listed in the Cross to Date column.

Source: Agrow Seed Company, *A Chronicle of Plant Variety Protection*.

In the early days of the development of the notion of PBRs, economic activities were seen largely as a product of industrial output. Industrialists reigned over the legislative process and helped give patent laws some of their present characteristics. A demarcation between patents and PBRs was introduced, largely to protect institutional empires and not necessarily to promote innovation. Current development, especially with the advent of biotechnology, show the weaknesses of the false dichotomy between patent and PBRs; the demarcation is being eroded. This erosion raises questions on the future of international institutions which, over the last two decades, have advocated the adoption of PBRs.

Debate about intellectual property protection of plants emerged in the

19th century. Interestingly enough, one of the earliest people to seek protection of his collection of plants was Pope Gregor IV, a contemporary of the Austrian monk, Gregor Mendel.[3] After the Paris Convention, plant breeders started pressing for the equivalent of patents in plant protection. These concerns were later consolidated into various laws, which were introduced in Europe in the 1920s, and the US in the 1930s, to protect new plant varieties. While the legislation accorded plant breeders intellectual property rights, it made clear the demarcation between biological and inanimate inventions so as to avoid the possibility for double protection.[4]

By the 1920s, some European countries were starting to shift their legislation to protect life forms. This is illustrated by the case of Germany. In 1922, the Reichsgericht (Supreme Court) upheld a patent protecting the use of a certain bacterium in the treatment of tuberculosis. In the early 1930s Germany started issuing patents for plants; this process was accompanied by seed registration, which had started in 1905. Seed registration was subsequently used as a medium through which varietal names could be used as trademarks, thereby conferring exclusive rights over breeding lines. Efforts to strengthen varietal protection were abandoned when the National Socialists came to power. Post-war attempts to reintroduce plant patent protection were not successful, as the government felt that such measures would inhibit the supply of high-quality seed.[5]

The history of PBRs is largely a result of competition between advocates of industrial protection and agricultural protection. While some countries pursued PBRs, others experimented with the provision of patent protection for plants – a provision offered in Germany, in the 1930s, for example. But this process met with strong political opposition: it was argued that plant patents would bring breeding to a halt and as a result, many breeders withdrew their application for patent protection. In addition, the procedural requirements for granting patents made it difficult for plant varieties to be protected. It was difficult for plants to meet the requirements for repeatability, industrial application and inventive step. In some cases, the patent law could not meet the scope of protection required for plant varieties.[6] In 1930, however, the US passed the Plant Patent Act, which granted patent-like rights for vegetatively produced plants.

The requirement for reproducibility proved a major obstacle for plant breeders. Their opponents argued that it was difficult for a person skilled in the art to reproduce a new variety without the person acting as an inventor or possessing special gifts. The objection, however, did not consider the fact that plant breeders were equipped with certain skills which enabled them to undertake their work. But the availability of certain skills alone was not enough to account for the fact that life forms are by nature difficult to reproduce. New plant varieties that result from mutations would be hard to reproduce, even if the mutations were induced by well-established procedures and skills.

The fact that breeders started with materials provided by nature made their claim to patent protection even more difficult. Their opponents argued that their products were a combination of natural and human activity and did not result solely from the creative power of the mind. This objection did not take into consideration the fact that the inventors of inanimate objects start with a set of existing innovations or base their work on intrinsic physical and chemical properties of natural material which are analogous to the genetic material that breeders start with. The functional units that constitute technological systems are a result of existing knowledge, key physical and chemical properties as well as human creativity.[7]

The final objection related to the requirement for the complete description of the invention. Inanimate objects are easier to describe because they do not change in time in a manner that would compromise their description. Such description is important because one of the requirements for getting patent protection is to disclose the relevant technical information so that a person skilled in the art can reproduce the invention. Since patents are supposed to help enrich the technical knowledge of a given society, such disclosure is necessary. It was argued, however, that adequate description could not be provided for plants and most of the information could be communicated only verbally.

In response to these obstacles, breeders sought protection under plant breeders' rights (PBRs) and these efforts continued during the post-World War II period. In France, breeders argued that property rights for new plants should be treated as a basic human right. Various European countries provided plant variety protection using different legal instruments. While the campaign continued, it became increasingly obvious in Europe that the protection provided hitherto was becoming difficult to administer and some lawyers were arguing against the validity of plant protection.

These national campaigns turned into international efforts to search for adequate protection for new plant varieties. At its 1952 congress in Vienna, the International Association for the Protection of Industrial Property (IAPIP) noted that it was necessary for new plant varieties to be protected either under patent law or by any other means. More specific suggestions were made by the International Association of Plant Breeders for the Protection of Plant Varieties (ASSINSEL) at its 1956 congress at Semmering, Austria. ASSINSEL stressed that there was a need for an international conference to study the matter at an official level and develop a convention that would provide the principles for new variety protection. By then, the movement towards new variety protection had become strong enough to turn the recommendation into concrete action.[8] The critical moment came in 1961 when the FAO launched the World Seeds Campaign and when a group of industrialized countries signed the International Convention for the Protection of New Plant Varieties (UPOV).[9]

UPOV thus represents the culmination of a long history of struggle by plant breeders to have exclusive rights over the products of their work. The convention aimed at introducing uniformity in plant protection law while at the same time allowing for variability in national legislation. It provided for protection either through PBRs or patents. PBRs, on the other hand, are in many respects analogues of patent laws. But unlike patent law, the convention also allowed for protection on the basis of discovery. This view was based on older legal interpretations which equated discovery with invention. Most of the modern laws do not grant protection for discoveries. In order to be granted protection, the new plant varieties must fulfil the requirements for distinctiveness, novelty, sufficient homogeneity, stability and varietal denomination.

The social implication of PBRs has been a subject of debate over the last decade. Some of the criticisms resemble those levelled against patents while others are specific to the agricultural sector. The first argument against PBRs is that they inhibit the exchange of germplasm. This view is based on the fact that the granting of monopoly rights by definition restricts the exchange of available materials, especially advanced breeding lines. It is, indeed, not expected that private breeders who want to put a particular variety on the market would allow for the free exchange of germplasm, especially to potential competitors such as farmers, public research institutes and other private breeders. This is a normal corporate response and would happen even without PBRs. What PBRs do is to legitimize the process and therefore grant breeders the right to refuse exchanging germplasm if this is likely to undermine their interests.

The issue, however, is not all that simple because germplasm exchange, especially among breeders or even competitors, will still continue. This is partly because it is a way of making germplasm available to other members of the industry, material that would otherwise be held by one person. The exchange among breeders is consistent with corporate behaviour. The analogue of this among industrialists is the exchange of technical information or patents. But such a system works only in conditions where the exchanging parties have similar interests. Even during fierce market competition, corporations are known to exchange technical information and this would apply to plant breeders. What is significant, however, is that PBRs offer the breeders an opportunity to be selective in the exchange process and to maintain access to exclusive benefits or monopoly profits.

The available information seems to show that the pattern of

information flow and genetic resource exchange has changed in recent years. While corporations have continued to acquire the needed material using a wide range of routes, the flow of germplasm from the private sector to public sector institutions seems to have been reduced in recent years. 'The amount of scientific information and plant breeding materials sent from universities to other universities and to private companies increased during the 1970s, but the flow from companies to universities decreased, perhaps in part as a result of the [Plant Variety Protection Act]. Given the increased value to private companies of scientific developments as a result of the [Act], private plant breeders would be expected to be more active in searching out and obtaining information and germplasm from public plant breeders.'[10]

Like in patents, the differential application of property protection among countries may lead to a restricted flow of technical information or genetic material. The history of the patent system has numerous cases where corporations have been reluctant to license their inventions in countries which do not have an effective patent system. Firms may also restrict the flow of information or genetic material to countries which are likely to allow the material to be marketed in violation of patent or PBR protection elsewhere. This would, therefore, tend to restrict germplasm exchange to UPOV members. The restrictions would not only guarantee monopoly rights, but would also help in putting pressure on other countries to sign the convention. The issue here is not necessarily the existence of PBRs, but the differential distribution of such rights and the capacity of the various countries to implement them.

The question that arises from this is whether, then, there would be more germplasm exchange without PBRs. In the first place, PBRs are products of certain political, economic and social practices which are consolidated under private property rights. To think of a situation where such rights would not be relevant also requires a vision of an alternative economic, political and social system. If such rights are withdrawn in countries which still have private property rights, then the breeders would tend to keep their varieties secret.[11] This would pose other practical problems. The fact that varieties are usually applied in the field and in some cases accessible to the public, would make the use of secrecy almost useless. Breeders may, in turn, tend to opt for hybrids which embody protective characteristics.

Another question that relates to PBRs is the narrowing of the genetic base for plant breeding. Opponents of PBRs argue that since breeding programmes aim at producing varieties that meet particular market and environmental requirements, the outcome is likely to be a narrower range of products because the solutions would tend to converge. This could lead to the introduction of varieties that in many respects are similar and could be vulnerable to similar pressures. This, however, should not be the case, because PBRs are allegedly granted for varieties that are distinct, which supposedly widens the product range. As indicated elsewhere, the nature of genetic resources makes it difficult to decide on distinctiveness, especially in cases where one negligible feature could lead to the granting of monopoly rights. New varieties may thus turn out to be no more than embodiments of cosmetic improvements.

The granting of monopoly rights tends to be associated with the concentration of the industry in question. The 1970 PVPA, for example, created suitable conditions for the acquisition of smaller firms by large corporations. But in order to understand why such processes are occurring, it is important to go beyond the limits of intellectual property rights and to examine the evolution of the industry and the major market and technological determinants of industrial restructuring. Intellectual property protection is largely a reflection of the co-evolution of technological change, industrial reorganization and institutional (as well as legal) reform. An examination of legal trends alone is likely to give a partial picture of more complex industrial structures.

The central concept that has helped to maintain PBRs as separate forms of intellectual property protection is the principle of demarcation between PBRs and patent protection.[12] Under this principle, most industrialized countries maintained separate exclusive rights for plant varieties and inanimate inventions. UPOV members were not required to grant patents for plant or animal varieties or essentially biological processes for the production of plants or animals. This provision, however, did not cover microbiological processes or their products. This UPOV provision is also reflected in the legal systems of various countries. The model laws proposed for Third World countries by the World Intellectual Property Organization (WIPO) adhere strictly to the demarcation principle.

The ability of UPOV to maintain the demarcation is weakened by a number of factors. Although UPOV Article 2(1) says that member states may not protect varieties by both patent and special rights, the Convention makes provisions for such protection to countries that provided both forms of protection prior to joining UPOV. This provision, in Article 37(1), was included as a condition for the US joining UPOV. This means that UPOV cannot prevent the US from protecting varieties by patents. The demarcation principle cannot hold in the US because the law has not been changed to eliminate the existence of any inconsistencies. Furthermore, the convention did not benefit from the advice and approval of the senate and so it does not carry the weight of a treaty. It does not, therefore, necessarily prevent the US from patenting plants.[13]

In any case, this demarcation is gradually being eroded by the introduction of a large number of biotechnology inventions in the economic system. This erosion is closely associated with the current efforts to reform legislation so that life forms are granted patent protection. This is probably one of the most controversial aspects of the growth of the biotechnology industry and the development of national as well as international legislation. The erosion of the demarcation not only illustrates the futility of the Cartesian dichotomy between industry and agriculture, but it also illustrates how technological change introduces new legal and institutional adaptations.

Patenting life forms

While most European countries shifted towards PBRs, the US Congress relaxed the requirements for description so that new plant varieties could be patented. The campaign for plant patent protection became a major preoccupation for breeders in the 1920s. Luther Burbank, who gave the US over 800 plant varieties, including the Burbank potato, was a leading advocate of plant variety protection.[14] He lamented:

> A man can patent a mousetrap or copyright a nasty song, but if he gives the world a new fruit that will add millions to the value of the Earth's annual harvest he will be fortunate if he is rewarded by so much as having his name connected with the result. Though the surface of plant experimentation has thus far only been scratched and there is so much immeasurably important work waiting to be done in this line, I would hesitate to advise a young man . . . to adopt plant breeding as a life work until [Congress] takes some action to protect his unquestioned right to some benefit from his achievements.[15]

His efforts were terminated by his death in 1926 but his widow continued the campaign until Congress passed the PPA in 1930, which allowed for the protection of asexually propagated plants.[16] To get around the requirement for novelty, the law required the new varieties to be distinct. An interpretation of the PPA provides protection for varieties which are distinguished by their habit, disease-resistance, soil condition, colour, flavour, productivity, perfume, form and ease of asexual propagation. Although the Act provided for patent-like protection, the legal requirements for granting the patents, such as non-obviousness, made the product space narrower than that accorded inanimate inventions.

Under the law, a farmer who reproduced trees from a patented variety would infringe the patent, but mutations or sports from patented plants could be reproduced and sold without infringing the original patent. Since 1930, the US has maintained two ways under which plants could be protected: through the asexual reproduction of 1) plants, and 2) seeds.

Asexually reproduced plants are genetically similar to their parents and are, therefore, viewed as distinct and uniform enough to qualify for patenting. Such plants have been protected under the PPA. This law was amended in 1953 to cover newly found plants in a cultivated state.

This law is basically an analogue of the regular patent law requiring that the plants be subject to the test of distinctiveness as an equivalent to the novelty requirement applied in regular patents. It prevents others from reproducing the same plant and therefore guarantees exclusive rights to the owner. The law, however, does not prohibit others from reproducing and selling mutants arising from a protected plant; for mutants are a deviation from the plant for which protection was provided; this is mainly because the protection is granted for plants that are distinctive. Plants reproduced through open pollination are not protected by patents because fertilization occurs by chance in a population with many heterozygous individuals. Since each plant reproduced in this form may be different, it is difficult to conform to the technical requirements for patent protection.

The use of inbred lines, however, yields patentable plants. Repeated self-pollination and selection of desired traits will lead to offsprings with two identical alleles in many of the gene pairs. Since the seeds produced thus also have the same genome, the inbred line will be relatively distinct, uniform and stable. This is the basis under which patent protection can be granted for plants under the Plant Variety Protection Act of 1970 (PVPA). Like the patent system, the Act provides breeders with limited exclusive rights on sexually reproduced plants in exchange for the disclosure of the information pertaining to the varieties and the methods of reproducing them.[17]

Fungi, bacteria and first generation hybrids are not covered by the PVPA. This does not directly apply to hybrid progenies. Seeds produced in this way will not breed true to type and the farmer cannot effectively replant the seed. This is a more reliable way of ensuring that the farmers rely on the breeder for seed for each crop; plant protection is therefore embodied in the seed. This form of plant protection is more effective than patent rights and has led to increased private investment in hybrids research.[18] Hybrids cannot be patented because they are not stable, but the PVPA allows control over the direct use of patentable inbred lines used in the production of hybrids. This guarantees that the hybrid is in turn effectively protected.

One of the requirements of the PVPA is that the new variety must be distinct from previous varieties by some set of descriptors. The vagueness in this requirement has opened up opportunities for granting protection on the basis of minor or cosmetic improvements. 'The effort now wasted in cosmetic breeding is significant. Varieties released by public experiment stations are backcrossed by private breeders to change some one or more of their descriptors such as chaff color in wheat and then patented and promoted as new varieties.'[19] One check against such cosmetic breeding is the rigid breeding standards maintained by public sector institutions, as well as the criteria used by farmers for adopting new varieties. As private breeders increase their activities, however, the chances of releasing varieties with minor improvements or changes are likely to increase.

Table 5.2
Crops with hybrid seed

Crop	Date of Seed Availability	Hybridization Method	%Area Under Hybrids (1980)
Corn	1926	CMS*/hand emasculation	99
Sugar beet	1945	CMS	95
Sorghum	1956	CMS	95
Spinach	1956	Dioecy	80
Sunflower	?	CMS	80
Broccoli	?	Self incompatibility	62
Onion	1944	CMS	60
Summer squash	?	Chemical sterilant	58
Cucumber	1961	Gynoecy	41
Cabbage	?	Self incompatibility	27
Carrot	1969	CMS	5
Cauliflower	?	Self incompatibility	4
Pepper	?	Hand emasculation	?
Tomato	1950	Hand emasculation	?
Barley	1970	Genetic male sterility	Negligible
Wheat	1974	CMS/chemical sterilant	Negligible

*Cytoplastic male sterility.

Source: Kloppenburg, *First the Seed*, p. 125.

Major advances in biotechnology have occurred in the last decade and these advances have put extensive pressure on existing laws, necessitating their reform.[20] One of the most significant developments in recent years has been the capacity to genetically modify life forms and use them in industrial applications. The modification of life forms through genetic engineering expands the possibilities for sexual and asexual reproduction thereby producing new plants or organisms with agricultural and industrial value. Industrialists and agriculturalists are thus seeking new ways to protect these innovations.

The current debates on the patenting of life forms tend to present the situation as if the process is a recent phenomenon. The patenting of life forms has co-evolved with other forms of intellectual property protection and what is happening now is a convergence of developments that have taken place differentially in other fields. In 1873, the US granted the first patent in the field of microbiology to Louis Pasteur. The patent included a claim to a biologically pure culture of a micro-organism, or a yeast culture as a composition of matter.[21] After the Pasteur patent, very few patents were granted for new micro-organisms,[22] partly because most micro-organisms occurred in nature and therefore could not be treated as inventions. The 'product of nature' principle dominated the legal scene following the *Ex parte Latimer* decision of 1889. In addition, most inventors were satisfied with process-type protection.

A new wave of inventions involving microbiological activity occurred in the 1940s, most of which were in the fermentation industry, especially for pharmaceutical products. In order to guarantee protection for some of these inventions, industrialists started the tradition of depositing micro-organisms with state authorities. The first such deposit was made in 1949 at the Northern Regional Research Laboratory of the US Department of Agriculture; the micro-organism deposited was for use in the manufacture of chlorotetracycline. In 1970 attempts were made to challenge the practice of depositing micro-organisms, but the US Court of Customs and Patent Appeals approved the practice. This ruling initiated a series of other court decisions which led to the famous *Diamond* v. *Chakrabarty* case in 1980,[23] which ended the long reign of the 'product of nature' principle.

The case started when the US Patent and Trademark Office (PTO) rejected a patent application for a bacterium of the *Pseudomonas* genus which contained at least two stable energy-generating plasmids that enabled it to break down multiple components of crude oil and could be used for digesting or cleaning oil slicks.[24] The micro-organism could not be protected as a trade secret because its use required release into the environment. The application was rejected on the account that the micro-organism was a living entity and, therefore, was not patentable. In a landmark ruling, the Supreme Court reversed the decision arguing that 'the patentee has produced a new bacterium with markedly different characteristics from any found in nature and one having the potential for significant utility. His discovery is not nature's handiwork, but his own; accordingly it is patentable subject matter under [the basic patent law].'[25] This decision became a major starting point in efforts to reform legislation to respond to the imperative of advances in biotechnology. Given the current advances in plant biotechnology, a large number of genetically-engineered plant varieties are expected to enter the market in the next decade. These plants will require patent-like protection.

Most of the major methods used in genetic engineering are protected under patents granted in 1980 and 1984 to Stanford University and the University of California (San Francisco) as a result of the work of Stanley

Cohen and Herbert Boyer. The two universities have monopoly rights up to 1997 over basic methods and tools used in the biotechnology industry. These patents are estimated to be worth US$1,000,000,000 in total. The two universities applied for patent protection but also allowed genetic engineering firms to use the process if they paid US$10,000 to the universities. The firms would get credits of US$50,000 against royalties claimed by the universities and pay only 50 per cent of the usual royalties on sales of genetically engineered products. Biotechnology firms have been paying the money, although Cetus, the second largest biotechnology firm, has voiced objection to the arrangement, arguing that the process has not led to major products and therefore the payment is taking away funds that could be used in further R&D.

Table 5.3
Projected release of new plant varieties

Crop	First Varieties Marketed	*In vitro* Gene Manipulation	First Transformed Whole Plants	Routine Growth of New Plants
Corn	Now	A few now	Early 1990s	Mid 1990s
Wheat	1984–86	1985–87	Early 1990s	Mid 1990s
Rice	Now	1985–87	Late 1980s	Early 1990s
Soybean	1988–90	A few now	Early 1990s	Mid 1990s
Tomato	Now	1984–86	1983–85	1986–88
Sugarcane	Now	1987–89	Early 1990s	Mid 1990s
Cotton	1983–85	1985–87	Early 1990s	Mid 1990s

Source: Sondahl, 'Applications for agriculture,' p. 15.

The US has begun to protect all plants under the utility product patent law. In September 1985, the Patent and Trademark Board of Patent Appeals and Interferences in the case of *Ex parte Hibberd et al.* decided that plant products were patentable even if they could be protected under the PVPA. The case involved an application for a utility patent for a maize variety with an increased content of tryptophan, an amino acid used in animal feed but found in low quantities in maize. The claims covered maize seed, maize plant tissue culture and maize plants developed by Molecular Genetics (Minnesota, USA). The decision was made on the basis that the statutes governing plant protection do not explicitly preclude plants from utility patent protection.[26] There is a range of options under which plants, in addition to other intellectual property rights, could be protected. The trend, however, is to bring most of the protection to fall under patenting while the other property rights will cover those inventions which cannot be protected under the patent law.

The 1980 ruling was followed by other legal reforms to facilitate the commercial application of scientific research. The Patent Act of 1980, for example, unified the various statutes and regulations governing the disposition of patent rights in federally-funded R&D projects. Under the legislation, the government retains non-exclusive, royalty-free licences as well as 'match-in' rights to inventions resulting from federally-funded projects. The legal reform was aimed at ensuring that the results of such projects are commercialized by the private sector without excessive bureaucratic obstacles. Such legislation would also minimize the potential competition between private- and public-sector R&D.

One of the potential impacts of the shift from PBRs to patent protection is a reduction in the flow of information among plant breeders. Under the PBR system, plant breeders were allowed to protect the product and not the process, which meant that the techniques available to plant breeders could be used widely by professional and amateur breeders. Under the patent system, however, scientists are allowed to protect the process and therefore limit the flow of technical information on the various methods of conducting research. The patenting of both products and processes enables the inventor to earn more royalty. The final costs may be passed on to the consumers since licence fees will be added to the final products. This suggests that society may in some cases pay more for products which result from the protected process.

These modifications in the legal regime have led to a significant stage where, since April 1987, the PTO now allows for the patenting of genetically-manipulated non-human multicellular living organisms, including animals. Although the decision does not cover human beings, it is possible that the law could be interpreted to include human traits. There are already antecedents to this problem. In 1976, John More learned that he was suffering from hairy cell leukemia and travelled to the University of California's Los Angeles Medical Centre for treatment. His spleen was removed and after surgery his symptoms abated, but for the next seven years he continued visiting the centre for blood tests; meanwhile, his doctor and technician had used a part of the spleen to develop a cell line that produced several potentially useful proteins.

In 1979 the university applied for a patent on the cell line, designated Mo, which was granted in 1984. The university has already entered into a collaborative project with a biotechnology firm for exclusive use of the Mo cell line. Although in 1983 More had signed a research consent form waiving all rights to any products of the university's research, later in the year he refused to sign a similar form. In 1984 he filed a lawsuit claiming that the university had misappropriated his cells and that he had a right to share the profits of any products derived from them. The case was dismissed in March 1986 and was later appealed.[27] This development already sets the pace for the patenting of human traits. Subsequent amendments of existing laws could also allow for the inclusion of human traits in patent protection.

Table 5.4
US utility patents and plant breeders' rights

	Plant Variety Protection Act	Plant Patent Act	Utility Patent
Complete written description required	No	No	Yes
Sexually reproduced plants protectible	Yes	No	Yes
Asexually reproduced plants protectible	No*	Yes	Yes
Protect hybrids	No	Yes	Yes
Novelty required	Yes	Yes	Yes
Requires standard of unobviousness	No	No	Yes
Provides generic coverage	No	No	Yes
Provides protection for genes and other parts of plants	No	No	Yes
Doctrines of Equivalents available	Unknown	Yes	Yes
Sexually reproducible varieties infringe	Yes	No	Yes
Asexually reproducible varieties infringe	Yes	Yes	Yes

* However, asexual reproduction infringes the certificate except when applicant is in pursuit of a plant patent.
Source: Williams, 'Utility Product Patent Protection,' p. 34.

The current activities to sequence the human genome are already raising intellectual property questions. The amount of money required to sequence the human genome runs into billions of dollars, which means that those who invest their resources in the exercise will expect reasonable returns. Since it is not clear who should own the genome or parts thereof, major legal and moral questions are likely to emerge in the next few years. The problem will be compounded by the fact that on-going legal reforms are making it more possible to patent human traits.[28]

It is feared that the patenting of animals will allow for the concentration of animal breeding activities in the hands of biotechnology firms; traditional breeders are likely to be displaced through the animal patent system. It is, however, still too early to adequately assess the impact of the new system. The patenting of animals is likely to affect the Third World countries in various ways. In the first place, the opportunities opened up by the legal reform will encourage the search for and collection of animal genetic resources in the Third World countries. Some of the on-going activities are likely to continue: in Kenya, for example, US scientists together with Kenyan researchers are trying to isolate the tick-resistant gene in indigenous cattle. The gene may be subsequently inserted into exotic high-yielding cattle which will then be able to survive in tick-infested areas. The introduction of such cattle may displace the indigenous stock, as they are likely to be raised on large ranches. If the tick-resistant animals are patented, it will be difficult for the local farmers to have access to them.

The US has initiated discussions with various Third World states – Pakistan is one – with a view to conducting research that would provide information on the safe flow of animal germplasm. (Discussions have also been held with Yugoslavia on this issue.) They include studies on the existence of pathogens in the semen of animals. New techniques are also being considered which enable researchers to determine the existence of certain pathogens directly instead of inferring them from the existence of antibodies in the animals. The collection of animal germplasm is also likely to be enhanced through the current techniques of embryo transfer. There is likely to be an increase in interest in the conservation of wildlife either in their natural habitat or in zoos. So far there are no internationally recognized standards in embryo transfer relating to germplasm viability or health requirements in international trade.

With the availability of the genetic material, researchers are able to breed new animal varieties by recombining genetic traits from other animals, humans, plants and micro-organisms. Already, pigs with a human growth hormones gene have been bred at the US Department of Agriculture's research station at Beltsville (Maryland). The only problem is that the pigs are cross-eyed and suffer from a variety of ailments, including arthritis. Researchers at the University of California (Davis) have fused a sheep with a goat, thereby producing a 'geep'. These R&D activities will not only produce patentable varieties, but will also lead to animals that cannot effectively reproduce and therefore the market will have to rely on the 'original' breeders, as in the case of hybrids.

The *Diamond* v. *Chakrabarty* case led to opportunities for increasing the range of coverage of protectible plants and plant materials as well as animals. Already there are increasing efforts to protect specific genes thereby ensuring a broader regime of protection for life forms and parts thereof. This trend makes it more difficult for Third World countries to assert control or ownership over the genetic resources from their territory because of the complexity involved in genetic manipulation. It would be beyond the scope of most Third World countries to establish the origin of particular genes that are expressed in an organism. This also undermines arguments that call for compensation to Third World countries for their genetic contribution to global agriculture.

It has often been argued that patenting life forms and seeds in industrialized countries will not affect production in the Third World countries because of the difficulties of enforcing those laws at an international level. These possibilities, however, are starting to emerge as the industrialized countries continue to strengthen their capacity to enforce intellectual property rights through international trade either unilaterally or collectively. It is estimated that the US loses US$8-20 billion annually on intellectual property infringement, of which South Korea accounts for US$70 million; the country has already agreed to revise its law to comply with US interests. Major changes are being considered for introduction in the operation of institutions such as the GATT which will strengthen its capacity to enforce intellectual property rights. In addition, several large corporations are considering not producing products that can be used in countries in which intellectual property law and adequate protection is absent.

These issues were placed on the GATT agenda in 1986 and definite agreements were reached on the subject. New rules and disciplines will be introduced in the GATT to facilitate the protection of intellectual property rights. The negotiations in the GATT will also aim at developing a multilateral framework of principles, rules and disciplines dealing with international trade in counterfeit goods; these goods will obviously include agricultural produce. It is important to recognize that Third World countries have little influence in the GATT and it is unlikely that new changes will favour their interest. The widening of the range of patentable invention in the US will create a situation in which the government could use patent law as a way of protecting the local industry; imports could be refused entry on the grounds that they infringe existing patents.

The changes will add to the already numerous protectionist barriers erected by the industrialized countries against goods from Third World countries. In addition to tariff barriers, industrial countries have adopted a wide range of non-tariff measures to control the flow of commodities from other countries;[29] there are hundreds of non-tariff barriers (NTBs) to international trade. NTBs are defined as all those public regulations and governmental practices that introduce unequal treatment between domestic and foreign goods of the same or similar production. The NTBs fall into five major categories: quantitative restrictions; non-tariff charges; government participation in trade; customs procedures and administrative practices; and technical barriers. These barriers have not only undermined the potential benefits to Third World countries from international trade, but have also raised major questions on the types of technologies and industrial adaptations these countries should introduce in order to compete favourably on the international market. The problem will get worse as more agricultural crops are affected by the NTBs, especially in view of the need in the industrialized countries to protect the nascent biotechnology industry.

These requirements are also likely to feature in bilateral agreements. This is exemplified by the case of the free trade agreement signed between the US and Canada in 1988. As part of the deal, the US has required Canada to reform its intellectual property law so that US investments and products can be protected. Canada is one of the few major industrialized countries without PBR legislation. Most of the plant breeding is still done by public sector institutions. There are several Canadian private and co-operative groups which are starting to undertake breeding and, therefore, see PBRs as a way of rewarding their efforts. These groups, in addition to the requirements of the free trade agreement, are pressing for the introduction of PBRs. Although in the short run such legislation may protect Canadian plant breeders, in the long run it is likely to lead to the control of the seed market by large US and European firms. In addition, Canadian firms will have to work much harder to be able to sell their seed in the US. Possibilities exist for the US to use intellectual property law as one of the NTBs against Canadian seed and other biotechnology products.

The need to reform legislation to conform to the imperatives of technological change is also being felt in Third World countries. But unlike the industrialized countries, which are setting the pace, the Third World countries require new legislation which reflects the emerging trends in the international scene. This is occurring at a time when there is growing interest in harmonizing the laws of the industrialized countries. Until Third World laws are harmonized, these countries may be forced to take unilateral initiatives before undertaking regional activities. Already, China has started working on its own legal framework to reflect its interest in biotechnology. Under the new law, which was promulgated in April 1985, inventions related to micro-organisms will now require the deposit of organisms before the patent can be granted.

China is not a signatory of the 1977 Budapest Treaty on the International Recognition of the Deposit of Microorganism for the Purposes of Patent Procedure, and cannot deposit its micro-organisms in other countries. This means that those deposited in China do not belong to a depository authority. Additionally, foreigners wanting to patent their inventions in China will be required to go to a patent agency designated by the government. Despite the fact that China is not a signatory of the Budapest Treaty, foreign firms may still want to patent their invention in

China because of the potentially large local market. In turn, China will use this bargaining power to acquire the technological and biological information related to the deposited micro-organisms.

Chinese law does not recognize product patents but patent may be granted for manufacturing and production processes. It also does not grant patents for foodstuffs, beverages, seasonings, drugs, plant and animal varieties, micro-organisms *per se* and substances obtained chemically or by using micro-organisms. China is currently sending its patent officers for training in the industrialized countries, and the European patent law is being used as the basis for reforming the existing legislation to reflect the country's long-term technological strategies. It is likely that subsequent reforms will bring the Chinese patent law closer to the European patent system.

African countries have not yet formulated effective policies to utilize their micro-organisms. Although there are various microbial projects, there is no reliable database providing information on the state of micro-organisms collected from African countries. The debate on germplasm has concentrated on plant genetic resources and has yet to deal with other fields, such as animal and microbiological resources.[30] African countries could improve their knowledge of the economic and environmental aspects of microbiological resources by increasing their participation in the MIRCENs. So far, these countries cannot effectively benefit from the MIRCENs partly because they lack the policies and institutional organization required to move into new biotechnology areas.

On the whole, developments in biotechnology are leading to major changes in the legal regime of various countries and these changes are likely to affect Third World countries in unpredictable ways. It is, however, possible to anticipate some of the effects by carefully examining the emerging trends and analysing the possible impacts. It is clear that the industrialized countries are already strengthening their collective capacity to enforce intellectual property rights, in addition to the measures already in place through trading practices. These reforms are taking place at a period of intensified conflict over the raw material for biotechnology – genetic resources.

Seed wars

Seed wars, as the international controversy over the ownership of germplasm and other related issues has been called, are partly a result of the success of the post-World War agricultural research programme with which the Green Revolution is associated. By the early 1970s the impacts of the revolution were starting to become more salient. One of the most obvious was the fact that germplasm freely collected from the Third

World countries was being sold back to them at high prices in the form of seed. In addition, the improved varieties could be used only with expensive inputs. The issue of genetic erosion also became more obvious to farmers as they could no longer find their indigenous varieties. The early 1970s also saw an increase in the flow of germplasm to the industrialized countries, especially after the establishment of the IBPGR. This period was also marked by increasing efforts by the UPOV secretariat to persuade more countries to sign the UPOV convention. The flow of genetic material from the IARCs to private breeders raised questions relating to the legal status of the material held by the centres.[31] In addition, the material held by IARCs was being used by private firms and protected by PBRs.

The introduction of semi-dwarf wheat varieties in the US was facilitated by the CIMMYT. The first major CIMMYT introductions were made in 1968 and by 1979 some 147 semi-dwarf varieties had been introduced in the US. Of these introductions, 18 were brought in directly from Mexico and 34 selected from Mexican crosses. Of the 95 selections made from US crosses, 14 contained Mexican varieties in their pedigree. By 1984, another 72 semi-dwarf varieties had been planted in the US, of which 25 contained genetic traits from CIMMYT or the Mexican national programme. The area planted by wheat containing germplasm from such sources was over 15.6 million acres in the mid-1980s and has been rising since then. The US agricultural economy has also benefited from rice, corn, soybeam, and cowpea germplasm originating from the IARCs.[32]

The early 1970s was also the period when the vulnerability of the global economy due to dependence on a few resources became more pronounced, especially after the 1973-74 oil price increases. The radical academic mood of the period was dominated by the 'dependency school' which cast international economic issues in a simple 'centre-periphery' model. Concern over genetic resources tended to fit well in this model. Here were resources predominantly located in the Third World countries being collected and improved upon by the industrialized countries and subsequently being used to exploit the Third World people. Like most other mechanistic models, this argument assumed that Third World countries were a homogeneous unit. They saw the process as a linear flow of benefits to the industrialized countries and costs to the Third World countries. It is in this context that the debate was started.

Discussions on the status of *ex situ* genetic resources have recently been a major topic at the FAO. The issue landed on the FAO agenda largely as a result of an international campaign conducted by the International Coalition for Development Action (ICDA) following the publication of Mooney's book, *Seeds of the Earth*. This culminated in a resolution passed at the 21st Session of the FAO Conference held in November 1981 calling for a legally binding convention on plant genetic resources and an international gene bank which would guarantee free exchange of genetic resources. The resolution was sponsored by Mexico, the home of the wheat Green Revolution.

Table 5.5
Area of US wheat lands occupied by varieties with germplasm from the world collection, 1984

State/Region	% Wheat Areas	Acres Planted
Hard Red Winter Wheat		
Kansas	50	6,817,500
Oklahoma	32	2,487,100
Colorado	22	828,400
Nebraska	16	515,200
Texas	10	740,000
Total	32	11,388,200
Soft Red Winter Wheat		
Indiana	30	351,000
Illinois	26	468,000
Ohio	15	186,000
Missouri	11	263,200
Total	19	1,268,200
Hard Red Spring Wheat		
California	100	770,000
Idaho	100	400,000
Washington	100	210,000
Oregon	100	80,000
Arizona	100	63,000
Utah	100	39,000
Nevada	100	16,000
Montana	16	347,600
South Dakota	15	255,000
Minnesota	14	308,000
North Dakota	8	430,550
Total	22	2,919,150

Source: Plucknett and Smith, 'Benefits of International Collaboration,' p. 52.

This resolution was discussed by FAO's Commission on Agriculture in March 1983 and was severely attacked by representatives from the industrialized countries. By then, the US had already indicated to FAO that it would not ratify any international convention on the subject. In order to avert a deadlock, FAO offered a compromise at the 22nd Session of the FAO Conference in November 1983. This was in the form of a voluntary

International Undertaking on Plant Genetic Resources which was not legally binding but aimed at the full exchange of genetic resources (including breeding lines and finished varieties). In addition, the conference also set up a special Commission on Plant Genetic Resources (CPGR).

The Undertaking was based on the view that plant genetic resources are a heritage of humanity and should be available without restriction. This 'common heritage principle' was developed by the Law of the Sea Conference and requires that no exploitation of specific resources be undertaken unless rules which guarantee equitable utilization are first established. Under this principle, management of the resources would be undertaken through an international body. These requirements would not strictly apply to genetic resources because they are already being utilized. Most of the elements of the Undertaking are in conflict with the requirements of private firms in general and PBRs in particular.

The effectiveness of the Undertaking has been the subject of speculation: 'The gap between the content of the Undertaking and present international practice, together with the opposition of developed countries and the seed industry to the Undertaking, make it doubtful, however, that international practice will change to conform to the Undertaking.'[33] The future of the Undertaking is still uncertain. Furthermore, it is not only its implementation that counts but also the public awareness that is generated and the possible national undertakings that might follow.

The Undertaking was signed by most Third World countries and some industrialized countries.[34] Since then, reservations have been presented to FAO by governments, especially on those sections which appear to be in conflict with national legislation and PBRs. Countries such as Argentina, Colombia, Cuba, Federal Republic of Germany, Jamaica, Mexico and Turkey have stated that their ability to make genetic resources available would be restricted according to their national laws or regulations.

Despite the reservations, the Commission on Plant Genetic Resources continued to work on the feasibility of setting up an international fund for plant genetic resources to bring the material collected through IBPGR under the auspices of FAO. The main issue of conflict has been the free exchange of all genetic resources (including breeding stocks and finished varieties). This is obviously in conflict with the national legislation of most countries. In addition, there have been discussions on how to involve countries which are not members of FAO – such as the Soviet Union – in the activities of the CPGR.[35].

At its 1987 meeting, the Commission established the International Fund for the Conservation and Utilization of Plant Genetic Resources. Governments and non-governmental organizations (NGOs) will contribute to the Fund on a voluntary basis and it will be administered as a Special Trust by the FAO. One of the issues to be discussed at later meetings will be the possibility of taxing seed companies a percentage of the sales value. This is expected to earn the Fund some US$150 million per year. Some private firms are expected not to be willing to accept the arrangements.

One of the key areas under discussion is the legal status of the material collected by IBPGR and located in various gene banks around the world. Spain, Costa Rica and Austria have declared that they would put their gene banks directly under the control of FAO. Also of interest is the legal status of *in situ* material. This is the most complicated aspect of the genetic resource question because *in situ* material is often considered as a public good and therefore claims of ownership are not taken seriously. One of the issues being considered for serious review relates to the question of farmers' rights as a way of asserting ownership over *in situ* genetic resources.

One of the most significant developments at the 1987 meeting of the CPGR was the recognition of the notion of farmers' rights. The meeting treated farmers' rights as being at par with PBRs, and the setting up of the Fund was partly aimed at creating an international institution that would act as a channel for compensating farmers for their centuries of work in genetic resource selection and conservation. It is the farmers' equivalent of plant breeders' royalties. This is an interesting feature because farmers' rights are not normally recognized in national legislation but they now have standing at the international level.

Although most African countries are signatories of the Undertaking, its elements have not been introduced into their legislation. Some of the countries already have legislation which grants PBRs. Kenya, for example, enacted the *Seed and Plant Varieties Act* which is largely dormant. It was introduced to:

> . . . confer power to regulate transactions in seeds, including provision for the testing and certification of seeds; for the establishment of an index of names of plant varieties; to empower the imposition of restriction on the introduction of new varieties; to control the importation of seeds; to authorize measures to prevent injurious cross-pollination; to provide for the grant of proprietary rights to persons breeding or discovering new varieties; to establish a Tribunal to hear appeals and other proceedings . . .[36]

The Kenyan law grants PBRs on the basis of breeding or discovery. It is not clear whether the granting of PBRs on the basis of discovery relates to plants brought into the country or those discovered locally. Plants are deemed discovered if they occur naturally in a form that makes them distinguishable by one or more important physiological, morphological or other characteristics from any other variety whose existence is a matter of common knowledge at the time of the application for a PBR. Common knowledge in this case refers to information on plant varieties under cultivation or other commercial use, or those included in recognized commercial or botanical reference collections, or those for which precise descriptions appear in any publication. From a strict interpretation of these requirements, PBRs may be granted for crops that are used in subsistence lifestyles but are not grown or known in the botanical literature. It is, therefore, possible that plants growing in the wild but used by local communities could be appropriated by individuals and protected by PBRs.

It is interesting to note that the law protects discovered plants while at the same time requiring that the varieties be pure, sufficiently uniform or homogeneous in their reproduction, and stable in their characteristics. These requirements are not easy to achieve in naturally occurring plants without further breeding. Most naturally occurring plants constantly undergo mutations leading to genetic drift and cannot easily be subjected to those preconditons. The protection is granted for 25, 18 or 15 years depending on the types of plants involved. The protection granted to breeders under this law would make it difficult to implement the Undertaking because of commercial interests. Introducing any of the elements of the Undertaking would require the participation of the private sector which is active in the field of breeding. But since the operating PBRs confer exclusive rights to the private sector, it is not likely that breeders would wish to sacrifice their private gains.

The central argument of the breeders is that it takes a long time and considerable financial resources before a variety is developed and put on the market. To ensure that the R&D expenses are recovered, it is necessary to guarantee that the new varieties reach the market and fetch a reasonable price. One way of ensuring that this happens is to restrict access to the material so as to maintain a certain degree of monopoly. As can be seen, this logic is in direct contradiction with the requirements of the Undertaking. Allowing the exchange of genetic resources under competitive conditions would tend to deter the private sector from investing in breeding activities since monopoly rights would not be guaranteed. It is, therefore, in the interest of the private sector to oppose any measures that would erode their capacity to maintain control over their competitive edge.

Although the future of the Undertaking is still unknown, it is expected that the obvious conflict between some of its provisions, especially on the free exchange of élite lines, is likely to meet with increasing objection from various countries. Many Third World countries already have laws that grant protection to improved varieties and they are not likely to relax the legislation and honour an Undertaking that is not legally binding. The conflict with national law is unlikely to be resolved in favour of the Undertaking without extensive international negotiations. The erosion of PBRs and the increasing role of GATT in international intellectual property issues are likely to complicate further the implementation of the Undertaking. It is also expected that some countries are likely to enter bilateral agreements under which they will honour each others' intellectual property rights.

The harmonization of intellectual property rights among industrialized countries will further strengthen the capacity of these nations to force Third World countries into honouring the PBRs and patents. With harmonized laws, it will also be possible to train Third World people in ways of enforcing intellectual property rights in conformity with the wishes of the industrialized countries. Currently, it requires patent officers in Third World countries to understand the variations in the laws of the various industrialized countries. The fact that some Third World countries are already adopting patent laws similar to those in the industrialized countries is likely to make any collective efforts by these countries even more difficult to achieve. This does not, however, rule out the possibility of unilateral action by some of the countries.

On the whole, the private ownership of genetic resources has been one of the most controversial aspects of international agriculture. The patenting of plants and animals, as well as their genes, has raised a number of moral and economic questions. Despite the objections, technological imperatives have made protection necessary, especially given the competitive nature of scientific research and the related corporate interests. As knowledge on certain techniques start to spread, so do the possibilities for producing technologies that belong to the state-of-the-art. This makes the need for intellectual property protection even more desirable.

Advances in biotechnology and the related institutional reforms have a wide range of implications for Third World countries. In the first place, the relocation of the production of some crops from the firm to the laboratory is likely to adversely affect the producers. Not only will these countries lose valuable foreign exchange earnings but also their positions on the international trade map will be revised. These changes are also likly to realign the geopolitics of international trade as the main sources of raw materials shift. Second, the patterns of access to genetic material and the related know-how are also going to change. There is likely to be a shift from collective bargaining, at least in the short run, to bilateral agreements. Third, the production of new plant varieties will be started in some Third World countries and thus change the existing relations of production. Fourth, the flow of technological information to Third World countries is likely to be restricted, especially in areas that are considered strategic. Finally, the direction of research in the industrialized countries seems to be inimical to the long-term interests of the Third World countries.

The capacity of some Third World countries to adapt to the changing conditions will depend on a large number of historical factors as well as the prevailing policies relating to genetic resources and biotechnology research. Some of the more advanced countries have the capacity to introduce some of the emerging technologies into their economies while others are still unaware of the potential contributions of biotechnology. The case of Kenya illustrates some of the problems that may face African countries wishing to strengthen their capacity in biotechnology R&D. Kenya, like many other African countries, did not benefit much from previous advances in technology such as micro-electronics. The country does, however, have the potential to move into biotechnology research, especially given its longstanding tradition in agricultural research. The next chapter traces the role of genetic resources in Kenya's history, and some of the institutional, mainly legal, obstacles that need to be removed before biotechnology can make any major contributions to the economy.

Notes

1. One could open and read such letters without breaking the seal. 'The first patent statute containing the major characteristics of contemporary patent laws is considered to have been enacted by the City State of Venice in 1474, the next landmark was Article 6 of the English Statute of Monopolies (1623). However, it was not until the advent of the Industrial Revolution that national patent laws became more widespread.' United Nations, p. 32.
2. Niklasson, 1987.
3. 'In the 1830's Pope Gregor IV declared property rights over all the plants in the Papal Gardens and in other gardens in the Vatican's dispersed, Italian lands. The attempt was not successful. In the 1860's, even as Gregor Mendel was writing down the results of his fantastic experiments, an American actually did obtain a patent of a process related to his fruit tree. And, in 1877, the German federation opened up the possibility of plant patents within industrial property legislation', Mooney, 1983, p. 137.
4. It will be shown later that this demarcation is no longer valid as technological imperatives are closing the gap between the various sciences and the resulting technological applications.
5. Bent et al., 1987, pp. 40-98.
6. Lange, 1985, p. 19.
7. The notion of functional units is explained in Clark and Juma, 1987.
8. By then, it was obvious that such initiatives could not be pursued through FAO, which had stopped working on the issues partly because a large number of governments were opposed to patent protection.
9. UPOV members include Belgium, Denmark, France, Germany, Hungary, Ireland, Israel, Italy, Japan, Netherlands, New Zealand, South Africa, Spain, Sweden, Switzerland, UK, USA.
10. CAST, 1985, p. 29.

11. Alternatively, agriculture may tend to become more secretive as farmers try to enforce secrecy. This is already being applied in some African countries such as Kenya where flower production, for example, takes place on 'fortress farms' with limited access to members of the public. Kenya grants PBRs but the legislation is largely inactive. In the absence of effective personnel to deal with infringement, farmers are forced to keep the production process and germplasm secret.

12. Some countries, such as Japan, do not restrict the protection of plant varieties under the patent law. In Japan, plant varieties 'are not excluded from patentability and, in addition, provisions establishing a preferential status for a patent right for a breeding process, which is not a patent right for a plant variety itself, are contained in the Seeds and Seedlings Law. However, patent rights for breeding processes are actually very rare and, furthermore, as these provisions are unreasonable in view of the legal nature of the rights, they must be interpreted and administered in a restrictive manner', Monya, 1985, p. 36.

13. Williams S. B. (Jnr.) 1986, p. 35.

14. One of the modern descendants of the Burbank potato is the Russet Burbank, the main source of McDonald's french fries.

15. Quoted in Witt, 1985, pp. 75-6.

16. Asexual reproduction is done through grafting, budding and tissue culture.

17. See Barton, 1982, for details.

18. 'Because hybrids expand the physical market for seed to the total acreage sown and because they make it possible entirely to disconnect seed prices from actual costs of production, they open up enormous opportunities for profit for private enterprise. In the last 15 years, most of the vegetable crops have become hybrid, but the great breakthrough that private (and, in many cases, public breeders as well) are pursuing is the production of hybrid seed for wheat, barley and soya beans', Berland and Lewontin, 1986, p. 788.

19. Schmid, 1985, p. 132.

20. See Plant, 1983, p. 95-105. See also Schmitt, 1983, and Adler, 1984. A historical survey of patenting in the field of microbiology is given in Irons and Sears, 1975.

21. US Patent No. 141,072. The claim reads in part: 'Yeast, free from organic disease, as an article of manufacture.'

22. The US granted the first patent for an antitoxic serum in 1877, for a bacteria vaccine in 1904 and for a viral vaccine in 1916.

23. Halluin, 1982, p. 68. This case built on a 1977 case (*In re Bergy*) which related to purified fermentate of vitamin B12. It is notable that the patent grant in this case was upheld despite a strong 'product of nature' argument and the fact that the product occurred in nature in impure form. The court's view was that the natural fermentate did not have as much medical value as the industrial product.

24. The plasmids provided a separate hydrocarbon degradative pathway.

The application was filed in 1972 by General Electric. For a review of other court cases, see Beier et al., 1985, pp. 100-105.

25. *Diamond* v. *Chakrabarty*, 447 US 303 (1980).

26. Arguments on the possibility for double or multiple protection do not hold in the US as the intellectual property law allows it in the case of design patent/copyright, design patent/trademark, and copyright/trademark interfaces. There are irreconcilable conflicts between utility patents and PVP certificates: '(1) The PTO acknowledges that new, useful and unobvious genes are eligible for utility patent protection . . . Therefore, a utility patent . . . would dominate a PVP certificate covering a variety containing that gene, and commercial use of variety without the patent owner's permission would constitute infringement. (2) Processes for reproducing and producing plants are eligible for utility patent protection . . . The use of a patented process to produce a variety covered by a PVP certificate would constitute patent infringement. (3) A trademark cannot be placed on a patented item and the combination commercialized without the permission of the patentee,' Williams, 1986, p. 39.

27. Tangley, 1987, p. 376. One of the questions being discussed is whether patients should be informed about the use of their cells in experiments that have potential commercial value.

28. For details, see Roberts, 1987a.

29. Deardorff and Stern, 1985, pp. 13-14.

30. See FAO, 1981b, for details on strategies for conservation and utilization.

31. Godden, 1984, pp. 215-18.

32. Plucknett and Smith, 1986b, pp. 52-4.

33. Bordwin, 1985, p. 1069.

34. By April 1987, the following countries had adhered to the Undertaking: Burkina Faso, Cameroon, Cape Verde, Central African Republic, Chad, Ivory Coast, Gabon, Guinea, Kenya, Liberia, Madagascar, Malawi, Mali, Mauritania, Mauritius, Mozambique, Senegal, Zambia, Zimbabwe, Bangladesh, Democratic People's Republic of Korea, China, Fiji, India, Republic of Korea, Nepal, New Zealand, Philippines, Solomon Islands, Sri Lanka, Tonga, Austria, Belgium, Bulgaria, Cyprus, Denmark, Finland, France, Federal Republic of Germany, Greece, Hungary, Iceland, Ireland, Israel, Liechtenstein, Netherlands, Norway, Poland, Spain, Sweden, Turkey, United Kingdom, Antigua and Bermuda, Argentina, Barbados, Bolivia, Chile, Colombia, Cuba, Dominica, El Salvador, Grenada, Haiti, Honduras, Jamaica, Mexico, Nicaragua, Panama, Paraguay, Peru, Bahrain, Egypt, Iran, Iraq, Kuwait, Lebanon, Libya, Oman, Syria, Tunisia and People's Democratic Republic of Yemen.

35. The Soviet Union and the German Democratic Republic participated at the 1987 meeting of the Commission and the Soviety Union was on the final drafting committee.

36. Republic of Kenya, 1977b, Cap. 326, p.4.

6. Germplasm and Kenya's Agriculture: A Case Study

Kenya is one of the few African countries with an advanced agricultural system. The country has accumulated extensive capability in agricultural research and it is possible to undertake major programmes to conserve the country's genetic resources while at the same time enhancing their utilization through the diversification of the food base. Kenya has the capacity to embark on an extensive biotechnology programme, but undertaking such programmes is likely to be inhibited by longstanding institutional and legal rigidities inherited from the country's colonial past. The programmes cannot be effectively introduced without major policy and institutional reforms to reflect the research and organizational imperatives of diversifying food production as well as conserving the genetic base over the long run.

Plant introductions

Kenya lies in a tropical region endowed with diverse ecological conditions ranging from the coastal to the alpine and associated with extensive genetic diversity. So far there has been no systematic study of the country's genetic diversity, although plant identification has been conducted in selected regions. Kenya lies close to the Ethiopia-Somalia centre of diversity identified by Vavilov. This region is the home of crops such as coffee, barley, castorbean (*Ricinus communis*), flax, lab lab (*Lablab purpureus*), onion, sesame (*Sesamum indicum*), sorghum, wheat, teff, millet, and okra (*Abelmoschus esculentus*). In addition to these well-known crops, Kenya is rich in other crops which form part of the traditional diet, including numerous species of peas (*Vigna* species), amaranth, spiderflower (*Gynandropsis gynandra*) and members of the Solanaceae and Cruciferae families. Large sections of the country are either arid or semi-arid and communities have, over the centuries, supported a wide range of drought-resistant sorghums and millets for various end-uses, ranging from food production to the fermentation of local brews.

The organization of local lifestyles in Kenya reflects the range of genetic diversity. The complexity of traditional socio-economic activities is largely based on the diversity of animals and plants which are used. The available ethnobotanical information on Kenya shows the interrelationships between social organization and genetic diversity. The introduction of new plants into the country and their integration into the socio-economic fabric has, however, dramatically changed the existing lifestyles and attitudes of many local communities towards indigenous genetic resources. This has partly contributed to the neglect of local genetic resources and the associated socio-cultural practices. These changes occurred hand in hand with the introduction of new institutions and knowledge systems which at times are inimical to the conservation of local resources. Although the case of conservation is today strong enough to justify major government and private initiatives, the dominant biases introduced during the colonial period continue to make it difficult for any major programmes to be initiated.

We have so far seen how the British transferred tea from China and subsequently eroded the monopoly of the Chinese in the tea trade. This crop was subsequently introduced in those British colonies with suitable conditions. The introduction of new crops in the colonies not only earned their economies revenue, but the fact that most of them were exotic guaranteed control over the production knowledge. Local labour could also be diverted from indigenous crop production to the new crops through regulation or coercion. From the early days of the evolution of Kenya's modern agriculture, there was a preference for exotic crops.

The early period of colonization guaranteed the availability of land for agricultural production, but the land was not useful unless accompanied by suitable genetic resources that could be effectively turned into agricultural yield. An 1897 report by the Commissioner and Consul of British East Africa, Sir A. Hardinge, lamented that traders and agricultural settlers were still outnumbered by officials, missionaries and railway employees. In 1901, Sir Charles Eliot (H.M. Commissioner) produced a report on the Protectorate which identified some of the crops that could be grown on the coastal strip. These included cinnamon, cardamon (*Elettaria cardamomum*), cocoa, vanilla, ceara rubber (*Manihot glaziovii*), sisal (*Agave sisalana*) and ramie. For Ukamba highlands he suggested European fruits, cereals, vegetables, tea, coffee, cotton and tobacco. Earlier attempts to introduce cattle from the coast and Somalia had failed and Charles Eliot recommended crossing local breeds with imported ones.

Already it can be seen at this early stage that the emergent agricultural system was building on exotic genetic resources. But Eliot also saw the possibility of extracting economic value from some of the local resources while at the same time introducing new genetic resources into the economy: 'The first necessity is to appoint a rudimentary Woods and Forests Department, which would examine the resources of the country and offer advice on such questions as the best means of preserving rubber and mangrove trees and the probability of success in cultivating coffee, tea, tobacco and vanilla.'[1]

Table 6.1
Area and values of selected commodities (1983/84)

Commodity	Area		Value (d)		Value per Ha	
	% of Total	Rank	% of Total	Rank	K£/ha	Rank
Milk	46.6	1	16.0	3.0	70	16
Maize and beans (a)	22.6	2	16.6	2.0	153	12
Root crops (b)	7.9	3	8.1	5.0	205	9
Sorghum and millet	6.7	4	1.5	11.0	48	17
Coffee	2.9	5	21.6	1.0	1,489	1
Wheat	2.2	6	2.1	10.0	191	10
Cotton	2.1	7	0.4	18.0	32	18
Fruits	2.1	8	3.1	9.0	296	7
Sugar	1.7	9	– 3.6	11.9	–432	19
Tea	1.6	10	11.9	4.0	1,325	2
Sisal	1.1	11	1.1	12.0	137	14
Vegetables	0.7	12	3.4	8.0	913	3
Cashew nuts	0.5	13	0.4	15.0	162	11
Groundnuts	0.4	14	0.2	20.0	84	15
Barley	0.3	15	0.4	17.0	249	8
Sunflower	0.2	16	0.2	19.0	141	13
Pyrethrum	0.2	17	0.4	16.0	419	6
Rice	0.2	18	0.5	13.0	519	5
Tobacco	0.1	19	0.5	13.0	885	4
Beef	(0)	—	6.8	6.0	(c)	—
Sheep and goats	(0)	—	4.9	7.0	(c)	—
Others	(0)	—	3.1	—	(c)	—
	100 (e)		100 (e)		170 (f)	

Notes:

(a) Because beans are typically interplanted with maize, the two crops are considered together; maize alone accounts for 13.3 per cent of total value.
(b) Includes potatoes, which accounts for 5.3 per cent of total value.
(c) No estimates available.
(d) Value at farm gate.
(e) The total area is 5.17 million hectares and total value is K£1,035 million.
(f) Excludes beef, sheep and goats, and 'others'.

Source: Republic of Kenya, *Economic Management for Growth,* p. 64.

In his second report (1902-1903) Eliot stressed the need to set up a large-scale Agricultural Department that would undertake research on the introduction of new varieties in Kenya. In a series of reports (1903-1904), the Director of Agriculture surveyed various parts of the country and recommended the types of crops that might be introduced there. These reports became an important source of information for the introduction of new crops. The current distribution of major commercial crops in Kenya is consistent with the recommendation of the report, except for the Kericho tea-growing zone which was not considered suitable for that crop. The omission of Kericho might have been due to perceived labour shortages in the area.

One of the outcomes of colonial agriculture was a consistent effort in plant breeding from which the current structure of the seed industry emerged. The structure of the industry has been influenced by the co-evolution of public and private sector breeding activities,[2] and the relationship between these two sectors has enabled the industry to expand to meet most of the needs of commercial agriculture. Government breeding work is organized under the Ministry of Agriculture, with the National Agricultural Research Stations (NARS), which are devoted to the major crops grown in the country, as the loci of activity. The NARS have their own networks of sub-stations which undertake research defined by the stations. The sub-stations that deal with specific crops include the Embu African Research Station, which breeds maize, and the Potato Research Station at Tigoni. Most of the maize released by the government is bred at the Kitale station. Most highland maize varieties are bred at Edebese/Kitale and Uasin Gishu areas on private farms or on the Agricultural Development Corporation (ADC) farms.

Maize varieties have been cultivated on the coast of Kenya since they were introduced in the 1500s by the Portuguese who probably brought them from the West Indies, although the centre of diversity for maize is Central America. There are several types of maize grown in Kenya which were introduced at different times. The popular East African Flat White originated in the US and was brought to Kenya at the turn of the century through South Africa; the Yellow arrived along the same route. The extensively variegated (purple, red and brown) Curzo maize was brought to Kenya from South America by missionaries in the early 1900s.

The seeds bred by the NARS are released after testing, certification and registration by the National Seed Quality Control Service (NSQC). This is a regulating institution whose subsidiary arms ensure that the new seeds conform to the rules and standards of the International Seed Traders Association (ISTA), which now has a Kenyan branch.[3] The major breeding activities are devoted to maize, wheat, barley, pasture and other commercial crops. Research on traditional crops is undertaken by government institutions, and commercial linkages with the private sector are not as well established as are those for the major commercial crops. The main source of commercial seed is the Kenya Seed Company (KSC) which operates as a parastatal organization, and is 52 per cent government owned (through the Agricultural Development Corporation). The other shareholders are the Kenya Grain Growers Co-operative Union (KGGCU) which holds 27 per cent of the shares and individuals who account for 21 per cent of the shares. As a parastatal organization, the KSC is charged with the responsibility of implementing those sections of the government policy which relate to seed development. Their activities, therefore, reflect to a large extent the degree to which government policies on this issue are implemented.

Table 6.2
Seed-producing organizations in Kenya

Company	Main Crops
Kenya Seed Company	Maize, small cereals, sunflower, rapeseed, grasses, sorghum
Njoro Seed Company	Wheat, barley
Agricultural Development Corporation	Potato
Hortiseed Kenya Company	Beans, horticultural crops
East African Seed Company	Beans, horticultural crops
Simpson and Whitelaw	Horticultural crops*
Ideal Seed Company	Beans, horticultural crops
Mount Kenya Agro-Industries	Potato, beans
Jardinage	Horticultural crops, beans
Kenya Highland Seed Company	Beans

* This firm was recently acquired by the Kenya Seed Company.

The KSC has devoted most of its resources to the development of seed for maize, wheat, barley, oats, triticale, forage, sunflower and grain legumes production. In order to increase its activities in the development of horticultural seed, the company recently acquired the Simpson and Whitelaw seed company; additional horticultural seed is also produced by Hortiseed Limited, a subsidiary of the KSC. The Hortiseed firm undertakes seed development while Simpson and Whitelaw is largely responsible for seed marketing. Some 75 per cent of all horticultural seed in Kenya is produced by Hortiseed from breeding material collected locally or imported. With the acquisition of Hortiseed, the KSC is gradually reducing its dependence on imported seed for local breeding.

This firm has not only been able to supply the local market with seed, but it has also started exporting seed to other countries in Eastern and Central Africa, as well as Denmark and Britain. In addition to these activities, the KSC is also starting to play a significant role as a base for seed multiplication for foreign firms, especially for export to Europe and North America.[4]

As indicated above, the modern agricultural sector is a product of the colonial economy which was largely based on exotic genetic resources. Over the years Kenya has continued to import as well as to collect local genetic resources for breeding purposes. Over the 1964-85 period, the country imported nearly 64 per cent of all the germplasm accessions used for breeding; cereals accounted for nearly 49 per cent of all accessions. The country is largely dependent on foreign sources for most of the major crops with nearly 88 per cent of the cereal accessions in storage having been imported. This should not be a surprise since the agricultural sector is based on exotic genetic resources. None of the major crops are indigenous to the country and, therefore, imports are necessary in order to maintain the breeding programme. Food crops in which the country has relatively large local accessions include cassava, sweet potato, pasture species and oil crops. None of these are major crops and their contribution to the economy is still marginal.

It is interesting to note that the local accession of cassava is larger than the imported one despite the fact the crop is indigenous to Brazil. This could be explained by the limited research underway in cassava improvement and the marginal interest in germplasm introduction. The crop is known to be relatively drought-resistant and provides flexibility in harvesting and storage and has, over the years, adapted to local conditions. Recent work at the International Institute for Tropical Agriculture (IITA) in Ibadan, Nigeria, has shown the potential of applying tissue culture techniques to cassava cultivation. It is equally notable that local or traditional accessions of vegetables are far fewer than imported sources; this also reflects the limited research devoted to local vegetables.

Similar trends can be noted in the accessions of forest species. Although local botanists have carried out extensive taxonomic work, only very limited effort has been devoted to the use of local species for afforestation and reforestation. Instead, imported species have dominated the government's forestry programmes. In 1985, nearly 95 per cent of Kenya's planted forests were exotic species, with cypress and pines (*Pinus* species) accounting for 78 per cent. Public perceptions, however, are starting to change as more local communities engage in reforestation activities. This has been enhanced by Presidential statements on the value of indigenous trees. These communities are more likely to build their programme on local knowledge and tree species. The challenge, however, is to create conditions that would enable these communities to work closely with the scientific community to improve the local varieties.

Table 6.3
Local and imported seed accessions, 1964-1985

CROPS	LOCAL	IMPORTED	TOTAL
Cash Crops			
Coffee	291	1,197	1,488
Tea	5,000	40	5,040
Others	162	120	282
Sub-total	5,453	1,357 (20%)	6,810
Cereals			
Wheat	840	7,000	7,840
Sorghum	1,015	4,100	5,115
Barley	45	3,600	3,645
Maize	350	2,696	3,046
Others	895	4,772	5,667
Sub-total	3,145	22,168 (88%)	25,313
Pulses			
Common bean	1,123	1,526	2,649
Others	532	1,721	2,253
Sub-total	1,655	3,247 (66%)	4,902
Roots and tubers			
Cassava	149	42	191
Sweet potato	32	18	50
Irish potato	28	670	698
Yam	—	3	3
Sub-total	209	733 (78%)	942
Others			
Pasture species	7,590	3,376	10,966
Oil crops	264	179	442
Sugarcane	104	519	623
Fibre crops	62	213	275
Vegetables	91	380	471
Fruit trees	114	363	477
Forest species	52	460	512
Sub-total	8,277	5,489 (66%)	13,766
GRAND TOTAL	18,739	32,994 (64%)	51,733

Source: Muturi, *Germplasm Conservation and Utilisation,* pp. 4-5.

A similar situation has long prevailed in the livestock sector. The composition of dairy germplasm at the Central Artificial Insemination

Station (CAIS) illustrates the narrow genetic base on which the country depends. Nearly 93 per cent of the germplasm produced by CAIS in the mid-1980s was from four exotic dairy breeds (Ayrshire, Friesian, Guernsey and Jersey). The local Boran accounts for only 2.0 per cent and the Zebu's share was only 0.2 per cent of the artificial insemination programme.[5] The reduction in the genetic diversity of local breeds was effectively undertaken in the colonial period through the setting up of artificial boundaries between various communities. The divisions were associated with administrative and legal measures that reduced the traditional exchange of livestock genetic material between the various communities. It should be noted that reduction in local herds had already been initiated before these measures were introduced. Large numbers of cattle were confiscated as part of the punitive measures used following the conquest of the local communities.

Table 6.4
Forest plantation area
(Hectares)

	1977	1978	1979	1980	1981	1982	1983	1984	1985	1986*
Indigenous										
Softwoods	8	8	8	8	8.2	9.4	9.7	9.7	5.0	4.6
Hardwoods	8	9	9	10	10.2	11.3	11.6	11.6	3.6	3.5
Exotic softwoods										
Cypress	51	51	51	53	55.1	57.5	59.2	62.7	70.0	71.4
Pines	60	61	62	62	63.5	65.2	67.2	68.3	61.5	70.7
Exotic hardwoods										
Timber	3	3	3	3	3.3	3.5	3.6	3.6	5.8	4.0
Fuel	7	7	7	7	7.1	8.3	8.5	9.1	14.6	13.5
Total	136	139	140	143	147.4	155.2	159.8	165.0	160.5	167.7

* Provisional.
Source: Central Bureau of Statistics, Nairobi.

In 1906, for example, nearly 12,000 head of Nandi cattle were confiscated and sold.[6] 'As soon as Africans had provided the raw material for the European stock industry great efforts were made to keep African and European animals apart. The basic principles of stock development in Kenya were to support European ranching and discourage African pastoralism.'[7] The control of the movement of local breeds was ensured through strict quarantine requirements which were widely instituted. In addition, the restrictions imposed by the fixed Reserve boundaries undermined the Maasai traditional methods of breeding, selection and germplasm exchange. The Boran 'bulls had always been selected with great care to produce animals adapted to dry conditions. The supply of such bulls came from Samburu and Somalia; with their fixed Reserve boundaries after 1912 and the European-settled lands forming a block to their north, the Maasai were completely cut off from new Boran stock.'[8]

Neglect of the local genetic pool therefore has a long history. Colonial agriculture and the reduction of the land available to local communities led to extensive soil degradation, which was partly associated with overgrazing. Systematic destocking of local cattle was introduced as part of the colonial programme to control overgrazing. The Athi River Meat Factory, for example, was established partly to help create a market for the by-products of the destocking programme.[9] In the late 1930s, cattle were confiscated from the Akamba and sold at prices that would make it profitable to the Factory.

This historical background not only condemned local livestock genetic resources to severe reduction, but also set in place institutional measures and arrangements which discriminated against their improvement. The subsequent emphasis on high-yield strains further marginalized the local stock from mainstream breeding programmes. Although there are efforts, mainly by development aid agencies, to study local animal genetic resources and conserve them, the efforts are not part of the central concerns of government activities in this field. Ironically, this is happening at a time when the industrialized countries are starting to show interest in the neglected animal genetic resources of the Third World countries.[10]

There are no data on the state of the local chicken genetic resources. This is partly because most of the breeding is done by the private sector and based on imported stock. The government, however, has made policy statements relating to the improvement of local breeds. 'There will be systematic improvement of the genetic potential of local birds, accompanied by training to improve management standards. The programme of cockerel exchange already being tried in Kenya will be evaluated and, if found suitable in genetic improvement, expanded.'[11] Another largely neglected area is that of fishery resources. Currently there are major questions being raised on the irreversible introduction of the Nile perch (*Lates niloticus*) which feeds on other fish in Lake Victoria. There has been a decline in certain species of *Haplochromis* and *Tilapia* in the lake which is attributed to the perch, overfishing and other factors. Kenya is currently involved in activities to introduce fish-farming in various parts of the country. The long-term viability of such efforts will depend on the capacity to conserve the genetic base of fishery resources for breeding programmes.

In some cases local genetic resources collected over the years have reportedly been lost due to poor storage conditions or neglect. For example, the Provincial Agricultural Research Station at Busia received large collections of improved cultivars and landraces of sorghum from other African countries (especially Uganda and Tanzania) for research and dissemination to farmers under the auspices of the East Africa Community (EAC). After the collapse of the EAC in 1977, the research stalled and the genetic resources were left unattended. By the time the Plant Quarantine Service (PQS) rescued the collection, some 50 per cent of the 80 sorghum varieties had been destroyed.[12]

The absence of effective *ex situ* preservation programmes in Kenya has also been a major obstacle to the conservation of genetic resources, and accessions have been reportedly lost due to poor gene bank management. In 1980, for example, the cooling equipment of the PQS seed bank at Muguga, Nairobi, broke down and was not repaired until 1983. Over the period, some 98 per cent of the accessions were destroyed. Of the 213 soyabean accessions, 183 were lost. Although some of the accessions can be replaced, the case illustrates the vulnerability of the current seed storage and management programmes. Ironically, this breakdown coincided with food shortages in the country.

In 1980, Kenya experienced serious food shortages, forcing the country to import maize, wheat and milk products; the shortfalls continued into 1981. The following year the country issued a food policy statement aimed at identifying measures for enhancing food security. The policy document stated that 'Kenya's agricultural development strategy is aimed at the continued expansion of productive investment, with the primary objective of the provision of basic needs and the alleviation of poverty through growth in agricultural output. The need to conserve national resources is now well recognized as an essential part of the strategy.'[13]

The document noted that the 'central objective of national food security policy is to ensure that an adequate supply of nutritionally balanced foods is available in all parts of the country at all times.'[14] The policy measures outlined include emphasis on drought-resistant crops such as sorghum and millet. The projections made for annual production growth rates required for self-sufficiency in sorghum and millet were, however, relatively low, possibly reflecting the anticipated low allocation of resources to these crops. Furthermore, there are no specific policy statements relating to germplasm conservation for long-term food security.

The national livestock development policy does not lay any specific emphasis on the diversification of the genetic base for the industry. It simply states: 'Breeding and selection, particularly wider use of artificial insemination and bull camps will be expanded.'[15] The policy focuses on increasing production using conventional breeding methods of selecting for output and not necessarily embodying genetic diversity into the programme. The food policy was more explicit on the need to diversify the genetic base: 'Livestock research will be directed towards improvement of the genetic potential of animals for arid and semi-arid areas and for zero and near-zero grazing systems.'[16] The explicit change in policy focus may partly have been prompted by the contemporary food shortages. Despite these statements, there is no major programme to conserve local genetic resources as a basis for guaranteeing the availability of material for improving the genetic potential of beef and dairy animals.

Table 6.5
Growth rates needed for food self-sufficiency

	Estimated Production ('000 tons)	Estimated Domestic Requirements ('000 tons)		Annual Production Growth Rate Required for Self-Sufficiency (%)	
	1980	1983	1989	1980-83	1980-89
1. Maize:					
(a) 1980 as base	1,620	2,777*	3.514	19.7	9.0
(b) 1976 as base	(2,264)**	2,777*	3,514	7.0	4.9
(c) Mean of 1976 and 1980 as base	(1,942)**	2,777*	3,514	12.7	6.8
2. Wheat Flour	142	292*	493	27.2	14.8
3. Sorghum/millet	369	445*	563	6.4	4.8
4. Rice	23	66	90	42.1	16.4
5. Beans	140	254	344	21.8	10.5
6. Potatoes	450	655	828	13.3	7.0
7. Sugar	4,402	342	571	−5.2	4.0
8. Beef	147	188	314	8.5	8.8
9. Milk***	1,259	1,615	2,058	8.7	5.6

* The figures exclude the production required to rebuild the strategic reserves.
** Hypothetical level of production.
*** Liquid milk and milk products expressed in whole milk equivalent.

Source: Republic of Kenya, *National Food Policy,* p. 49.

The need to divert attention to non-conventional crops is also reflected in recent development plans. The focus, however, will continue to be on the dominant crops: 'Agricultural research policy will focus on the development of technologies appropriate to the arid and semi-arid lands and on development of labour-intensive technologies appropriate to smallholder food production. This will be achieved through work on the introduction of new crop varieties . . . While the main emphasis of the plans is on increasing food production on small farms, attention is also given to other crops.'[17]

By emphasizing increased food production as the main focus of breeding programmes, Kenya has tended to drift towards a narrower genetic base in major commercial food crops. This, however, does not reflect the preferences of the consumers, who tend to opt for greater variability in their food sources. The case of beans illustrates this point; beans originated in Central and South America and reached Europe in the 16th century from where they were spread to the coast of Africa by Portuguese merchants. Since their introduction in East Africa, probably 300 years ago, beans have undergone significant variation in size, shape and colour. This process has partly been aided by consumer preferences.

A study conducted in the late 1970s showed that consumers preferred to maintain diversity in beans by not discriminating between any varieties on the basis of colours, sizes or shapes. The functional characteristics selected included not only nutritional value and yield, but the consumers selected some bean types for colour and taste. 'It was . . . learnt from farmers and extension staff . . . that most if not all bean types grown are acceptable, even light coloured small-seeded ones and that for instance the medium-sized Red haricot beans are highly esteemed in Embu District as they colour the food they are cooked with attractively.'[18] From a sample of 997 beans, the study concluded that of 'the ten main seed types identified, the Rose cocos were widely spread . . . (36.5%) but their colours and variegation patterns, sizes and shapes, differed considerably. The Canadian wonder types were second (13.1%) and varied from red-brown to dark purple and from medium to large in size . . . The Red haricot was widely spread and came third in frequency of occurrence (9.2%). The Mwezi moja was fourth with 9.0% . . . Its colour varied from light purple to dark purple-grey.'[19]

This consumer and producer approach to food diversity differs considerably from the preferred strategies used in industrial production, where the tendency is to move towards monocultural production and uniformity. Commercially induced 'diversity', for example in colour, is added during the processing of the food. This tradition seems to have a long history; the spice trade was partly a result of efforts to introduce 'diversity' in otherwise monotonous diets. Food colouring is a big industry in the developed countries and has recently been a major target for regulatory intervention. The importance attached to diversity within species is related to the need to conserve genetic resources. There is even a stronger case to conserve genetic resources among species. Most of the traditional lifestyles depend largely on natural diversity for food security; recent studies have recorded the use of over 53 wild species for food in southern Turkana (in an arid area covering 9,500 km^2 to the south and west of Lake Turkana).[20] This information has been uncovered through ethnobotanical surveys. Despite its significance, this is not the kind of knowledge that regularly informs policy formulation.

Table 6.6
Ten common bean types in a 997-seed sample

Local Name	Language	Description	Occurrence Frequency (%)
Tongmire	Luo	Rose coco, variegated red-brown on pink, medium-sized, oval	10.4
Wairimu	Kikuyu	Red haricot, small – medium-sized, oblong	9.2
Mwezi moja Kibuu	Swahili Kikuyu	Mwezi moja, with many fine purple spots, spots, medium – large-sized, oblong	9.0
Nyamariu	Kikuyu	Rose coco, variegated red on cream, medium-sized, globular	8.7
Gituru } Gitune }	Kikuyu Kikuyu	Canadian wonder, purple-black, medium – large-sized, oblong	7.0
Kathiga	Meru	Rose coco, variegated purple on cream, medium-sized, oblong	7.0
Kabumbu	Kikuyu	Canadian wonder, red-brown, medium – large-sized, oblong	6.1
Mwitemania	Kikuyu	Mwitemania, resembling pinto bean, variegated green-grey on cream, small – medium-sized, globular	2.4
—	—	Rose coco, variegated with large red flecks on cream, medium--large-sized oblong	2.4

Source: van Rheenen, 'Diversity of Food Beans in Kenya,' p.453.

There is thus a disjunction between policy statements and reality. In the first place, implementing these policy statements would require major changes in the current institutional arrangements and resource allocation. Second, the bias in the training of manpower cannot be adjusted over the short-term. Another major obstacle to crop diversification is the philosophy that underlies some of the country's economic policies. The promotion of alternative food crops or animal breeds needs institutional support to ensure their development and establishment in the market. The philosophy that guides the policies is, however, often inimical to institutional support for new innovations. It has tended, under the influence of the World Bank and the International Monetary Fund (IMF), to emphasize the efficacy of market mechanisms at a period when innovations are still too nascent to be subjected to the strong selecting pressures of supply and demand or factor proportions.

The livestock development policy document, for example, states that the 'production and marketing systems will be made more efficient through the forces of competition in the industry. The Ministry will intervene only where essential functions and services are not delivered adequately.'[21] This approach would tend to select out those breeds which have not been established in the market. On the whole, it is notable that the country needs to make deliberate efforts not only to formulate new policies, but also to introduce alternative institutional arrangements that would guarantee the conservation of genetic resources and their sustainable utilization. There is a need to examine in detail the existing institutions and identify the various obstacles to genetic resource conservation for guaranteeing long-term food security.

Colonial law and genetic resources

The historical evolution of Kenyan laws relating to natural resources in general, and genetic material in particular, was inimical to the rise of a conservation ethic in the country. Instead, a strong utilitarian ethic underpinned the development of the law. This was partly because the laws were formulated to facilitate the conversion of sections of the country into a colonial economy. It was also at a time when some natural resources, such as forests, were seen as an obstacle to development and had to be cleared to give way to modern economic activities. A parallel development in natural resource legislation reduced the potential for public participation in the management of the resources.

While the commercial use of forests was being promoted, efforts were undertaken to protect forests from local use. The isolation of forests was given legal standing and therefore new forms of land-use emerged under which access to natural resources was restricted. The early forestry laws were aimed at ensuring that the railways had easy access to wood for fuel. Under the *Ukamba Woods and Forests Regulations* of 1899, wood within two miles of the railway could not be cut without the permission of the District Officer or other person appointed by the Sub-Commissioner of Ukamba. The regulations were replaced in 1902 by the *East Africa Forestry Regulations*, which were broader but did not change the exclusive rights conferred upon the railways. Subsequent changes in the legislation enhanced the utilitarian ethic and focused on the administrative measures which ensured that forest resources were used for commercial purposes.

The introduction of protected forests at the turn of the century was the first move to alienate the indigenous populations from the resources in their locality. It should be noted that the areas declared forests could subsequently be turned into agricultural or commercial land by the Governor. Forests could also give way to public facilities such as schools or health centres. It appears, therefore, that the protection of forests was aimed at ensuring that the resource and the land on which they stood could be made available for selected human use.

As colonial agriculture expanded, problems associated with soil management started to emerge. These concerns later filtered into the legislation governing the management of forestry resources. It is notable that the utilitarian focus had not changed; forestry conservation was undertaken partly to protect agricultural land from degradation. The forestry policy that emerged in the 1950s emphasized the maintenance and improvement of climatic and physical conditions, conservation and regulation of water resources, and preservation of soil and its fertility. The overall objective was to improve the national economy. The need to undertake conservation, especially of the soils, had emerged in the 1920s and 1930s when the limitations of imported agricultural practices were starting to show.

> The Kenya settler, whether of farming stock and experienced in the vagaries of the English climate or an optimistic ex-service man . . . was at a loss when he had to deal with the heat, the droughts, and the torrential downpours of rain which are frequent in most parts of Kenya. There was little evidence of any appreciation by the farmer of the need to nurse his soil and to adapt his farming methods to the soil and the climate. Square fields were remembered in England; square fields must be right in Kenya. A dead straight furrow was the sign of a first-class ploughman; it must be good in Kenya.[22]

The concern over soil degradation led to an international search for knowledge, forcing agricultural officers to visit the US in the late 1930s to study their practices and introduce them into Kenya. These efforts led to major changes in the conservation policy, and a national programme was launched in 1946. Soil conservation, which was previously restricted to settler agriculture, was later extended to cover the so-called African reserves, and coercive methods were introduced to ensure the implementation of the programme. These changes in the policy regime did not, however, take into consideration the need to conserve genetic resources for expanding the local food base. Where conservation was deemed relevant, it was to guarantee the exploitation of the indigenous resources for short-run economic growth. Indeed, colonial agriculture was partly aimed at introducing new plants and reducing the amount of time and land devoted to indigenous crops. In part this was because the prevalence of indigenous crops was seen as a source of competition for labour with the commercial crops.

Although the colonial government made some effort to promote the production of indigenous commodities, especially where they contributed to commercial production without competing with the settler produce, there were strict regulations governing the commercialization of the so-called 'African produce'. The crops included local legumes, sorghums, potatoes, rice, millets, wheat, fresh vegetables, fresh fruit, onions and simsim grown by Africans.[23] *The Crop Production and Livestock Act* of 1926 introduced restrictions on the transportation, storage, commercialization and pricing of the crops. These restrictive laws were accompanied by the aggressive introduction of exotic plants and their promotion through research and administrative measures. The need to promote particular crops was subsequently consolidated under the *Agriculture Act* in a section which empowers the minister to declare certain plants 'special crops'.[24] This section provides for extensive administrative and financial support as well as the development of manpower exclusively for the promotion of these special crops.

Under the provision, the post-colonial government has since declared pineapples, wheat and sugar-cane (for the production of white sugar) to be special crops. The plants declared special crops after 1967 include apple, pear, peach, nectarine (*Prunus persica* var. *nucipersica*), quince (*Cydonia oblonga*), plum, apricot, citrus, avocado, paw paw, guava, mango, loquat (*Eriobotrya japonica*), white sapote (*Casimiroa edulis*), litchi (*Litchi chinensis*), passion fruit (*Passiflora edulis*), banana, plantain, strawberries (*Fragaria* species), cape gooseberry (*Physalis peruviana*), mulberry, berries, date palm, custard-apple (*Annona reticulata*), melon and grape. Macadamia nuts (*Macadamia tetraphylla*), pistachio (*Pistacia vera*) and oysternuts (*Telfairia occidentalis*) as well as geranium (*Geranium* species) and salivia oil plants were later added to the list. In order to promote horticultural production, the government also declared a wide range of vegetables special crops.[25] Most of the plants declared special crops and, therefore, allocated extensive financial and scientific resources, were exotic to the country or had already been incorporated into commercial agriculture; no comparable provisions were made for local species used in subsistence agriculture. This, coupled with the colonial restrictions over indigenous crops marginalized the local genetic resources from mainstream food production.

Conservation in modern Kenya

One of the main laws governing genetic resources is the *Forests Act*, which was first introduced in 1942 as the consolidation of the various pieces of legislation and regulations governing the use of forest resources.[26] As pointed out previously, the colonial resource policy was primarily based on the need to exclusively exploit the natural resources. Conservation measures were later introduced where it was felt that the exploitation could endanger future development prospects. The Act covers Central Forests, forests and forest areas in Nairobi area and on unalienated government land. The Act empowers the minister responsible for forest management to declare a forest area or a Central Forest or any part thereof to be a nature reserve for the purpose of preserving the constituent natural amenities and its flora and fauna. The minister can also revoke the protection. The law prohibits the cutting, taking, injuring, grazing and removal of forest produce or the disturbance of flora without the permission of the Chief Conservator of Forests. Such permission is restricted to conservation of the natural flora and amenities of the reserve.

This is a significant provision because it also restricts the commercial utilization of any forest products. Hunting and fishing are prohibited in the reserves except where considered necessary for purposes of resource management. Because of the difficulties for the authorities to ascertain the source of any forest product, the law requires the suspect to prove that the product is not from a protected forest. This implies that anybody found with a forest product is presumed guilty until proved innocent.

The legal provisions governing nature reserves have their analogue in the *Wildlife (Conservation and Management) Act* introduced in 1976. Under the Act, the director of wildlife management may set aside a section of a national park for use as a breeding place for animals or as nurseries for plants.[27] The Act also provides for the protection of animals and vegetation in areas adjacent to national parks, national reserves or local sanctuaries. This provision has significant implications for genetic resource conservation since it can be used to protect areas that do not fall under the strict jurisdiction of the wildlife and forestry legislation.

The protection of genetic resources could also be invoked, albeit indirectly, under the *Antiquities and Monuments Act*, introduced in 1983,[28] because the Act states that the declaration of a monument could include 'a specified site on which a buried monument or object of archaeological or palaeontological interest exists or is believed to exist, and a specified area of land adjoining it which is in [the archaeologist's or palaeontologist's] opinion required for maintenance thereof . . .'[29] Such a declaration would therefore limit access to the area for development, agriculture and livestock. No attempt has so far been made to protect genetic resources under this Act although its invocation would extend the protection of genetic resources to areas that are currently not covered by other laws. Since the management of monuments falls under the National Museums Board, it would require complicated co-ordination between the various ministries to effect the invocation, especially in the absence of legislation specific to genetic resources. Moreover, such invocation would be seen as encroachment by the National Museums Board on the jurisdictional terrain of the Forestry Department.

The protection of plants under the current legislation is also provided by the *Grass Fires Act* introduced in 1942.[30] This law requires that permission be granted by the authorities before anyone burns vegetation which is not his property, and thus lays more emphasis on property than on resource conservation. Although enforcing the legislation may indirectly lead to the conservation of plants, the Act does not make any specific reference to resource conservation.

The protection of local plants, especially against diseases and pests, is provided in the *Plant Protection Act* introduced in 1937,[31] which aims at protecting local plants against those pests or diseases whose effects are deemed by the minister responsible to be difficult to control or eradicate. It is also under this law that most matters pertaining to the introduction of exotic species are handled, especially through the quarantine services. The law also prohibits the export of plants that are likely to spread disease or pests. There are, however, no provisions prohibiting the export of genetic resources as a way of protecting the local material. Given the importance of agriculture in the Kenyan economy the penalty for the wilful introduction of pest or disease is surprisingly lenient. Under the law, any 'person who knowingly introduces any pest or disease into any cultivated land shall be guilty of an offence and liable to a fine not exceeding two thousand shillings or to imprisonment for a term not exceeding six months.'[32] Given the fact that large sections of Kenya's crops are based on monocultures, it is perfectly possible to wipe out large parts of the agricultural sector with the introduction of a specific pest or disease.

The impact of new pests, especially where there is little knowledge on the organism, could have far-reaching effects, as a recent Tanzanian example illustrates. In 1979 Tabora farmers noticed the existence of a rare pest which was causing extensive damage to stored grain. The grain borer, *Prostephanus truncatus*, was traced to Central America although it is still unknown how the pest reached Tanzania. The pest attacked grain that was stored in traditional ways and was therefore mainly a problem for subsistence farmers. The grain destroyed by the pest was valued at US$87 million in 1985.[33]

It is notable that sections of the current legislation may lead to the loss of genetic resources whose value is still unknown. One of these is the *Suppression of Noxious Weeds Act* introduced in 1945,[34] which was introduced to protect agricultural crops from weeds. Its powers extend from local weeds to those affecting the whole country. It is presumed that a plant shall be declared a weed only after it has been studied and proven so; this is, of course, in relation to agricultural crops. No provisions are made to assess other potential benefits of such plants. So far darnel (*Lolium temulentum*), jimson weed (*Datura stramonium*), *Datura ferox, Datura tatula*, water hyacinth (*Eichhornia crassipes*), giant salvinia

(*Salvinia molesta*), animate oat (*Avena sterilis*) and wild oat (*Avena fatua*) have been declared weeds.

One of these weeds, water hyacinth, is now being studied in different parts of the world as a source of numerous industrial products, including paper. It can also be a source of biomass for the production of biogas. Indeed, the plant was declared a weed without any considerations of other end uses. The prolific nature of some weeds makes them suitable for large-scale propagation. Some plants that are often considered weeds in some countries have recently become a major source of genetic material for reforestation and soil reclamation. It would be suitable for the law to include alternative end uses for the weeds instead of merely condemning them to destruction.

Recent activities in Kenya show that this approach could be incorporated into the existing legal framework as exemplified by efforts to control salvinia, a weed that threatens the ecosystem and fishing activities of Lake Naivasha.[35] In 1969, the government, under the *Fish Industry Act*, introduced the Fish Industry (Lake Naivasha) (Salvinia) Regulations, aimed at protecting the lake and the fishery resources against the weed. The regulations alone were not adequate and the government has more recently been looking into other ways of reducing the salvinia problem.

In 1983, the National Task Force for Biological Control of salvinia was launched by the National Environment and Human Settlements Secretariat of the Ministry of Environment and Natural Resources. One of the objectives of the group was to explore the possible utilization of salvinia. Analyses conducted by the National Agricultural Laboratories (NAL) showed that the weed could be used as animal feed, and that its nutrient and mineral contents were comparable to grass hay or maize silage. The weed also showed micro-nutrient levels comparable to other sources of organic manure. Moreover, dry salvinia is brittle and breaks down easily, therefore integrating itself into the soil more freely.[36] This clearly indicates the possibility for the economic use of the weed. With the advent of biotechnology, some of the plants, such as wild oats, now treated as weeds may contain genetic material that can be used in breeding and other sectors of the economy.

As can be noted from the above review, the current legislation does not specifically commit itself to protecting the country's genetic resources as a way of guaranteeing long-term food security. The laws carry a strong utilitarian ethic and were formulated at a time when the resource base was viewed largely as a source of economic benefits. Moreover, the restrictive regulations that were introduced at the time, coupled with the policy of colonial agricultural expansion, were inimical to the conservation of genetic resources used for subsistence agriculture. In contrast, traditional agriculture was based on the concept of diversity and long-term risk-reduction. The limitations in the legislation have now led to a search for alternative ways of protecting genetic resources.

The inability of the current legislation to offer effective protection for the country's genetic resources has led policy makers and scientists to search for alternative approaches. One of these has been the protection of specific plants or animals by Presidential decree. In November 1986, the President of Kenya, Daniel arap Moi, declared the *Aloe* a protected plant. The declaration was made on the basis of Presidential powers and not any specific legislation that pertains to plant protection. This was the first time in Kenya's recent history that a long-term conservation effort had been directed at a specific plant.[37]

The protection of the *Aloe* was prompted by extensive exploitation of the plant for commercial purposes, as well as the destruction of its habitat. The *Aloe* will now be exploited through plantations which would be established in the semi-arid parts of the country. There are at least 275 *Aloe* species and new plants are still being discovered all over the world. The *Aloe* has in recent years been a target of scientific enquiry and commercial interest because of its medicinal properties; its extracts are currently used in medicines and cosmetics.

Species of the *Aloe* which were originally collected from Kenya have been subjected to scientific screening for medicinal properties at the Jodrell Laboratory of Kew Gardens. The plant was reported to be promising and the Jodrell Laboratory hoped to turn it over to the private sector for commercial exploitation. The plant was collected from Kenya with the help of local scientists. It is not known how much of the technical information pertaining to the commercial value of the plant has been shared with Kenyan researchers.[38] The plant is widely used as a source of traditional medicine in many countries; it is this type of information that attracted Kew collectors.[39]

The Presidential declaration shows clearly that there is a need for specific legislation to protect the country's genetic resources.[40]

In the absence of legislation, the discovery of commercially useful species may easily threaten the future availability of such plants. Like the *Aloe*, wild species may suffer from overharvesting, and this may pose a major threat if the plant has a low reproductive rate. This is exemplified by the harvesting of the plant locally known as Mudziadyah (*Maytenus buchananii*) in the 1970s. More than 27.2 tonnes of the shrub were collected by the US National Cancer Institute from a game reserve in the Shimba hills for testing under a major screening programme that involved the collection of plant material from different parts of the world.[41] This plant, which is used in folk medicine to treat cancerous conditions, yields maytansine which was considered a potential treatment for pancreatic cancer. The first harvesting was done in 1972. When additional material was needed in 1976, it was found that the plant had a low reproductive rate and the region had to be searched for additional material. Efforts to locate it in Tanzania were fruitless. 'Thus, in order to mitigate the depletion of wild *Maytenus* populations in the reserve, a more careful effort was made in 1976 to collect the vinelike shrubs from a different Shimba Hills population.'[42]

That the plant was available at all could partly be attributed to the provisions for plant protection under the wildlife protection law. If the plant had been located outside a protected area, there would have been no way of enforcing any measures for sustainable utilization or harvesting. This case illustrates the need for legislation to guide the harvesting of plants for research programmes, especially in cases where their reproductive capacity is unknown and the plant may be threatened. This would not only guarantee that the plant would be available for further use in the same programme, but it would also ensure its availability to other research activities and continued local use.

For any legislation to be effective, more systematic research will have to be undertaken on the state of Kenya's genetic resources so as to establish the baseline on which legislation will operate. Without such studies, legislation will be ineffective and it will be equally difficult to provide the required institutional support for genetic resource conservation. Moreover, in the absence of information on the resource base it will be difficult to identify which are the critical areas that require special attention. The Presidential decree should, therefore, form the beginning of new efforts to understand the country's genetic resources. Not only is it necessary to understand the extent of genetic diversity, but it is also important to have information on the state of Kenya's germplasm in international or foreign gene banks.

Kenya and the global gene banks

The issue of accessibility to genetic resources that have been collected in one country and preserved in another has become a major issue of international debate. Kenya is one of the Third World countries from which an extensive collection of genetic resources has been made. The plants collected under the auspices of IBPGR include sorghum, millet, pasture crops and rice.[43] Some of the genetic material collected in Kenya has been deposited in public gene banks in various countries.

The genetic resources held in gene banks are usually the wild and weedy ancestral species and primitive cultivars; these are the source of variability from which the modern or commercial plants have been developed and the material whose legal status is currently disputed. According to IBPGR, '[n]o specific legislation exists conferring proprietary rights to any of these . . . categories and the basic genetic resources of crops remain "free" for distribution and utilization.'[44] The report admits that the 'transfer of material under development may be more difficult to obtain than hitherto.'[45] We are told that '[n]ot only the genebanks supported by the [IBPGR], but all reputable collections subscribe to the principle that the material which they conserve, and the data which characterize it, shall be freely available to all *bona fide* breeders and research workers.'[46] The issue is, however, whether countries have any legal right of access to genetic resources collected from their territories.

Kenya does not have any legislation governing plant genetic resources collected from its territory and deposited in foreign gene banks. Most countries with gene banks have no legislation that relates specifically to the legal status of the genetic resource held therein. Any analysis of the legal standards pertaining to the collection, ownership, deposit or maintenance obligations, and accessibility would, therefore, have to be based on the legal instruments of establishment, statutes and bye-laws of specific gene banks. Genetic resources are collected through various means: they may be collected directly by representatives of a gene bank as part of their programmes; gathered in response to specific requests; received through exchange programmes; or through unsolicited donations. Genetic resources may also come from other gene banks, institutes or private individuals. Expeditions to collect genetic resources are usually conducted through consultation with national governments.

Collections made under the auspices of the IBPGR have been undertaken in consultation with the Kenyan government, and copies of the collected samples have been left with the Kenya Agricultural Research Station at Kitale, the University of Nairobi, the Ministry of Agriculture at Muguga or the Sorghum and Millet Department Project of FAO at Lanet or the National Gene Bank in Nairobi. Until the mid-1980s there was no specific central co-ordination of genetic resource collection efforts or a central gene bank in the country, although various agricultural research stations have had their own collections for breeding purposes. The material is under government control and the legislation pertaining to national property would apply to them. What is of interest, however, is the legal status of material that has been collected from Kenya and deposited in foreign gene banks.

Concern over the legal status of genetic resources held in foreign gene banks resulted from reports in the 1970s that some countries or firms were restricting the flow of exchange of genetic resources for political and commercial reasons. In a widely quoted 1977 letter from the US Department of Agriculture, the administrator of the Agriculture Research Service informed IBPGR that although the country exchanged genetic resources with other countries, '[p]olitical considerations have at times dictated exclusion of a few countries.'[47] Surveys conducted in the early 1980s showed an increase in the number of restrictive practices in both developing and industrialized countries. An Irish semi-governmental body allowed exchange only among UPOV members, and India allowed the exchange of black pepper only within the country.

Private firms also restricted the exchange of genetic resources or, as in the case of banana germplasm held by the United Brands Company in Honduras, did not make available information pertaining to the collected resources. Ethiopia placed an embargo on coffee germplasm exchange in 1977, in response to which Ethiopia was informed by IBPGR that the country could not expect external collaboration in genetic resources activities if it did not make coffee genetic resources freely available for international exchange.[48]

Not only is sovereignty being extended to the collected genetic resources, but corporate secrecy is also making it more difficult for firms to exchange material that may be deemed of significant economic value. In many countries, sovereignty or national security and commercial interest coincide. Indonesia and Malaysia have been reported to be reluctant to share mango genetic resources with other countries. Other restrictive techniques have also been reported. For example, Taiwan would make sugar-cane resources available only ten years after the release of a variety, in order to prevent competitors from introducing varieties based on the same genetic resources.

The world's millets are kept at ICRISAT, NSSL, and at the All India Coordinated Millet Improvement Programme (AICMIP), Poona, India; Office de la Recherche Scientifique Outre-Mer (ORSTOM), Bondy, France; Plant Gene Resources office (PGR), Ottawa, Canada; and Crop Germplasm Institute (CGI), Beijing, China. Sorghums are kept at ICRISAT, NSSL, All-Union Institute of Plant Industry (VIR), Leningrad, USSR; Plant Genetic Resources Centre (PGRC), Addis Ababa, Ethiopia; Research Institute for Cereals and Technical Plants (RICTP), Fundulea, Romania; Sugar Crop Field Station (SCFS), Meridian, USA; Mayaguez Institute of Tropical Agriculture (MITA), Mayaguez, Puerto Rico; and American Sorghum Project (ASP), Tihama, Yemen. Kenya's sorghum and millet collections are deposited at ICRISAT and the NSSL. Recent rice collections have been deposited at IRRI.

ICRISAT was established in 1978 under an agreement between the Indian government and the Ford Foundation (acting on behalf of CGIAR) as an international, non-profit making, autonomous and philanthropic organization; it has full legal personality. The Indian government has extended the operation of Clause 3 of the United Nations (Privileges and Immunities) Act of 1947 to ICRISAT, thereby making it an international organization. Although ICRISAT has no clear statement on the ownership of the genetic resources held therein, it considers itself 'the owner of the plant genetic resources which it has collected or received.'[49] This view, however, is held by the senior officials of the organization and it is not certain that a court of law would uphold it. The situation would be more complicated if the dispute involved national governments, as matters of diplomacy would step in.

ICRISAT distributes genetic resources to scientists (in the private and public sectors), national or international gene banks at no cost and there are no legal provisions restricting this exchange. It is, therefore, generally understood that countries such as Kenya can obtain genetic resources back from these institutions under the normal provisions of free scientific exchange. There are, however, no legal provisions that would enable Kenya to recover its seeds purely on the grounds of origin. It can be argued that without prior agreement on the status of the material, ICRISAT's claim to ownership would hold and Kenya would not have any legal access to the material.

IRRI, like ICRISAT, is an autonomous, non-profit and philanthropic organization which enjoys privileges and immunities accorded by the government of the Philippines. IRRI's policies are formulated by its 15 corporate members of whom the Minister of Agriculture, the President of the University of the Philippines, and the Director General of IRRI are members. The rest of the members come from the international community (mainly rice-producing countries and donor institutions). Three of them are members-at-large and are elected with the approval or concurrence of CGIAR.

IRRI sees itself as a custodian or depository of the genetic resources in its bank. Clearly, this is a vague claim because it does not ascertain the degree of legal ownership over the material. 'In that context, however, it is not clear on behalf of what legal persons the material is held and whether the institute's freedom to dispose of such material is limited by the rights retained by third parties.'[50] This is, of course, assuming that the countries of origin make it clear that they still retain ownership of the material and IRRI shall serve only as a custodian. The position of a custodian is clear only if the terms under which the material is held is agreed upon by the two parties. Normally, a custodian would be appointed by the country of origin with the clear understanding that the material belonged to the source and would not be used, or exchanged or disposed of without consent. The practice of gene banks declaring themselves custodians is largely that of legal fiction and does not establish the status of the material held therein. Moreover, access to base collection may be restricted on grounds of low stock.

The fate of germplasm originating from Kenya is also made uncertain by the current trend in some industrialized countries to sell public sector institutions to the private sector. For a long time Kenya's germplasm has been deposited at the gene bank of the Plant Breeding Institute (PBI) at Cambridge, UK; Kenya was assured of access to the material because of the longstanding scientific collaboration with the UK. In 1987, the gene bank was sold to Unilever, the largest food corporation in the world. Although the fate of the gene bank is still unknown, it is expected that some of the material will be taken over by Unilever. This raises crucial questions regarding the ability of industrialized country institutions to safeguard genetic resources originating from other countries. In a period when state policies are reducing funding to public sector institutions and at times selling them off, it may not be wise for the Third World countries to entrust their material to these countries. The PBI will serve as a warning to other countries and possibly stimulate them to establish their own storage facilities.

A slightly different situation applies to the material held by the gene bank at Kew. Like IRRI, Kew considers itself as a custodian of the material held in its base collection. The material is supposedly held in trust for the scientific community for the benefit of humanity. But the material is owned by Kew unless it is deposited with specific conditions to the contrary. As far as is known, no Kenyan material has been deposited to any foreign gene bank with specific instructions asserting Kenyan ownership.

The legal status of material in national gene banks is clearer than that held by IARCs. The material is the property of the state and subject to the national legislation of the country in which the gene bank is located. This not only applies to direct government gene banks but also to banks that belong to institutions owned or controlled by the government. This holds for Kenyan material held at the NSSL; Kenya has no legal claim to this material on the basis of origin as the seeds now belong to the US government. Similar conditions apply to the material held at other government institutions or research stations. Kenya's material deposited in foreign national gene banks is now the property of those countries and the legal status of those in foreign IARCs is still unclear.

On the whole, Kenya has developed a large pool of skills that could be mobilized to initiate a wide range of biotechnology programmes. What is lacking is the awareness and political will to establish the kinds of institutional arrangements that would facilitate the implementation of various projects. The country has already initiated policy reforms in the educational sector which could help in the long-term accumulation of technological capability. The challenge, however, is to formulate explicit policies that would guide the country towards new levels of advancement in biotechnology and the related fields.

The case of Kenya illustrates how historical developments have shaped the direction of agricultural R&D. Any efforts to introduce biotechnological innovations into the economic system will require major changes in the existing policies, legislation and institutional arrangements. One of the features of most African countries is the inability to reorganize institutions and reform legislation fast enough to respond to the imperatives of technological changes. If these changes are not made, the enormous opportunities provided by biotechnology will be missed and the potential impact of developments in other countries will not be avoided. It is notable that despite the challenges, Kenya has been able to introduce a number of policies which support conservation efforts. What is needed is a major re-evaluation that would lead to reforms in science and technology policy as well as genetic resource utilization and conservation.

Notes

1. Luckham, 1959, pp. 97-105.

2. The history of the plant breeding in Kenya is crop-specific, except for efforts which were associated with particular policy requirements. For example, World War II led to policy requirements for self-sufficiency in temperate vegetables. This was a requirement from the British government and new measures as well as personnel were employed to facilitate the introduction of temperate vegetables. The activities included the establishment of experiments and field trials at Njoro. Some of the work was sub-contracted to farmers who grew seed for the Agricultural Department for distribution in the other British East African colonies. For details, see Hawkins, 1944.

3. The procedure also involves the inclusion of the new varieties in the national Index. Kenya also abides by the principles and procedures of seed certification prepared by the Paris-based Organisation for Economic Cooperation and Development (OECD) which is used by the major industrialized countries.

4. KENGO, 1986, p. 48.

5. Muturi, 1986, p. 9.

6. Zwanenberg with King, 1975, p. 92.

7. To enforce this, extensive powers were conferred to the administrative officers under the *Crop Production and Livestock Act* first introduced in 1926 to regulate the availability and reproduction of local breeds. The reduction of local breeds was also effected by restrictions on grazing rights. Governor Belfield emphasized the colonial view: 'I deprecate in the strongest possible way the suggestion that pernicious pastoral proclivities should be encouraged by the grant of any rights for grazing purposes. My policy is to discourage the proclivities by every legitimate means, not only because they are productive of nomadic tendencies but because they inculcate in the minds of people a distaste for any settler industry,' Ibid., p. 92-3.

8. Ibid., p. 94. As an evolutionary process, breeding and selection are more effective under open systems. The colonial system, however, was 'closed' and therefore reduced the rate and possibilities for genetic exchange between the various pastoral communities. This was reinforced by the administrative and legal measures introduced to encourage the development of the settler livestock industry which was largely based on imported genetic material.

9. The overgrazing was partly a result of the displacement of the Akamba

during the colonial period to create room for the introduction of cash crops such as coffee. The ensuing pressure on marginal lands led to overgrazing and land degradation.

10. See, for example, National Academy of Sciences, 1981 and 1983.
11. Republic of Kenya, 1980b, p. 31.
12. KENGO, 1986, p. 28.
13. Republic of Kenya, 1981, p. 1.
14. Ibid., p. 19.
15. Ibid.
16. Ibid.
17. Republic of Kenya, 1983b, pp. 182, 183.
18. Rheenen, 1979, p. 449.
19. Ibid., p. 452. A total of 526 samples were collected from farms in Machakos, Embu, Kisii, Kakamega and Kiambu while the rest were collected from agricultural shows in Nairobi, Kakamega and Embu, making a total of 997 samples from Kenya's main bean-growing areas. It is notable that the main seed types also vary according to their time of maturation. The Canadian wonder is a late maturing type common in Central Province, the Rose coco is a medium maturing cultivar of Eastern, Central and Western Provinces, and Mwezi moja is an early maturing cultivar used in the drier areas of Machakos and Kitui Districts.
20. Morgan, T., 1981, p. 101.
21. Republic of Kenya, 1980b, p. 17. Emphasis on market mechanisms are likely to skew the allocation of resources in favour of established production systems and therefore undercut the potential for promoting alternative options. Indeed, the evaluation techniques used under such conditions lead to absolutist solutions such as dependence on one or a few dominant sources of agricultural produce.
22. Maher, 1950, pp. 45-55.
23. Republic of Kenya, 1977c, *Crop Production and Livestock Act,* Cap. 321.
24. Republic of Kenya, 1980a, Cap. 318, Section 190.
25. Special crops for horticultural development include tomatoes, carrots, Brussels sprouts, cabbages, cauliflower, lettuce, potatoes, egg plant (*Solanum melongena*), okra (*Abelmoschus esculentus*), celery, cucumbers, onions, garlic, spinach, chillies, sweet pepper, green beans, French beans, peas, pumpkins, asparagus, keralla, cowpeas, cluster bean, courgettes, marrows (*Cucurbita pepo*), dioscorea, dudi, globe artichokes (*Cynara scolymus*), kohl rabi (*Brassica oleracea* var. *gongylodes*), leek, loofah (*Luffa aegyptiaca*), mushroom, New Zealand spinach (*Tetragenia tetragonioides*) and green maize. All flowers and decorative plants as well as miraa or khat (*Catha edulis*) are special crops.
26. Republic of Kenya, 1982, Cap. 385.
27. Republic of Kenya, 1977a, Cap. 379. This Act replaced the *Royal National Parks of Kenya Ordinance* of 1945 which had similar provisions.
28. Republic of Kenya, 1984, Cap. 315.
29. Ibid., Section 4(1)(*b*).
30. Republic of Kenya, 1972, Cap. 327.
31. Republic of Kenya, 1979, Cap. 324.

32. Ibid.

33. 'The final conclusion from the story of the grain borer's arrival in Africa is that it could happen again, with new pests. The movement of vast amounts of food and seed grain around the world means that nowhere can regard itself safe from pests that exist elsewhere. And the main burden falls on the poor', Cross, 1985, p. 11.

34. Republic of Kenya, 1983a, Cap. 325.

35. *Salvinia molesta*, formerly called *S. auriculata*, is a native of South-Eastern Brazil. The plant is kept under control in South America by various predators such as beetles (*Crytobagus singularis*), moths (*Samea multiplicalis*) and grasshoppers (*Paulinia acuminata*). The plant was introduced in Kenya by settlers in the 1930s to adorn their aquaria, and grown in ornamental pools and aquaria in Nairobi in 1936. It later started to grow in the Nairobi River and subsequently in the Athi River. The weed showed up in a dam in Kitale on Nzoia River in 1953 and the dam had to be drained and dried to kill it. It was first reported in Lake Naivasha in 1962. In two years the plant had covered over 60 hectares south-east of the Crescent Island. The area was quarantined and twice sprayed with the herbicide Gramoxine S that year. Another spraying was done in 1965 along the western shore of the lake. In the meantime, a hidden salvinia population was building up to the north-eastern part of the lake, covering up to 3.0 square kilometres by 1970. The population was concealed behind masses of papyrus (*Cyperus papyrus*). A storm later ripped apart the papyrus and released the *Salvinia* into the rest of the lake. By the early 1980s, the weed was estimated to cover up to 20 per cent of the lake. Efforts to introduce the Trinidadian grasshopper, *P. acuminata* failed, as the predator could not be established due to low night temperatures and predation by local birds. Aerial spraying in 1979 was stopped because of concern over the threat of the herbicide to nearby horticultural crops. Earlier sprays had noticeably affected other aquatic flora.

36. Kanani, 1983, pp. 20-21.

37. The first time such a decree was instituted was in 1915 when the Governor of Kenya used the powers conferred upon him by the *Forest Ordinance* of 1911 to declare the bark mangrove (*Rhizophora mucoronata*) a protected tree. This decree was issued after regulatory efforts (*Mangrove Regulation* of 1900) to reduce the rate of bark mangrove harvesting had failed.

38. Over the same period, Kew Gardens also collected from Kenya new types of coffee in Wajir Garissa area, Taita-Taveta and Shimba Hills. There is considerable interest in the Taita-Taveta collection because the coffee has large beans but is not currently planted commercially. The characteristics could either be transferred to the currently utilized plants or it could be developed as a new plant for commercial use.

39. 'During the time of Alexander the Great, *Aloe* was cultivated for use as a purgative. In folklore medicine, it has often been used for inflammations of the skin and eyes, and for sores, minor cuts and burns. In 1935 its efficacy against X-ray burns was demonstrated. In modern-times aloin extracts and powdered *Aloe* latex has been used as a purgative, especially for chronic constipation. Today it is used mainly as an ingredient in tincture of benzoin, which capitalizes on its antibacterial and skin-healing properties', Oldfield, 1984, p. 97.

40. Previous attempts to use decrees include a Presidential declaration making it illegal to cut down trees.

41. The collection was part of a programme by the National Cancer Institute to screen plant 'families of special interest' (FOSI) which included Apocynaceae, Celastraceae, Compositae, Simarousbaceae, Rutaceae, and Thymelaeceae as well as the families that make up the order of Magnoliales. The *Maytenus* genus belong to the Celastraceae family. Screening plants for pharmaceutical properties is becoming a major research agenda for large drug firms and agencies such as the European Economic Community. Screening projects are currently underway in countries such as Indonesia, Mauritius and Zimbabwe.

42. Oldfield, 1984, p. 137.

43. The legal status of IBPGR itself has been questioned. According to the European parliament, the 'creation of IBPGR . . . has meant that control of crops and conservation has passed from the hands of the member governments of the United Nations into those of an ill-defined group of governments, scientists and private foundations from the developed countries . . . the IBPGR lies outside any form of international control which has enabled genetic resource material to be transported from the centres of origin to the gene banks of developed countries', Anon, 1986, p. 25.

44. IBPGR, 1983, pp. 6-7.

45. We are assured that if for some reason genetic resources are eroded, 'the conservation of collected germplasm in genebanks ensures the availability of variation for further crop improvements', ibid., p. 6 and p. 8.

46. Ibid., p. 8.

47. Letter from T. W. Edminister, Administrator, Agricultural Research Service, US Department of Agriculture to Richard Demuth, Chairman, International Board for Plant Genetic Resources, Rome. The US has at various times reportedly restricted germplasm exchange with countries such as the Soviet Union, Afghanistan, Albania, Cuba, Iran, Libya, and Nicaragua. Usually, restrictions on scientific exchange, unless specified, include the exchange of germplasm and the related technical knowledge.

48. IBPGR, 1980.

49. FAO, 1986a, p. 22.

50. Ibid.

7. The Way Ahead: Policy Options for Africa

This exploration has shown that the world has entered a new phase in which innovations in biotechnology are likely to play a major role in socio-economic evolution. This is a period of major agricultural and industrial discontinuities requiring national and regional policy changes as well as the enhancement of technological capability. These changes are likely to affect African countries in various ways. Unlike previous technological revolutions, biotechnology offers a wide range of opportunities for Africa. Not only is biotechnology suitable for diverse and decentralized economic activities, it is also amenable to participatory research and can be controlled at community level. Moreover, the fact that biotechnology is science-intensive reduces the capital-related entry barriers. Successful programmes in biotechnology will depend largely on national and regional policies on scientific and technological development as well as genetic resource conservation.

Evolution of economic systems

Policies are conceived in an epistemological vacuum. The Aristotelian worldview and its subsequent refinements have given the world an agricultural and production model that is inimical to long-term requirements for socio-economic evolution. The world has already started searching for heuristic methods and metaphor on which to build a new worldview. Never before have new findings in the natural sciences been subjected to the philosophical scrutiny as advances in quantum physics have provided. The Cartesian-Newtonian worldview is under constant attack and some of the leading contributors to the new paradigm are beginning to demonstrate the implications of their work for everyday life.[1] It appears that economics, on which development policies are based, is being left behind again. While the scientists were rejecting the Newtonian metaphors last century, economists were busy basing their theories on these very notions. As a result, economics is one of the most moribund of the disciplines; but economics is so important that it cannot be entrusted in the hands of a few priests whose main aim is to maintain the old order.

African countries are entering the biotechnology era at a time when major academic traditions are being challenged and a new wave of questioning is starting. These are interesting times marked by a measure of chaos in epistemology and this should form the basis for rethinking Africa's future. There is an urgent need to re-examine the epistemological underpinning of the prevailing economic theories and the resulting policies in light of the continent's needs. The prevailing economic policies manifest numerous features that seem inconsistent with Africa's long-term interests. In the first place, the transformation of the continent will take a relatively long time – several generations. This requires policies which are based on long-run economics. The Keynesian view that in the long-run we are all dead does not make sense for underdeveloped countries which are faced with major internal and external development problems. Socio-economic evolution is about the long-run; it is about time and irreversibility. Neither can Africa rely on neo-classical economics, which view the long-run mainly in terms of allocative efficiency of scarce resources.

Secondly, traditional theories have led African countries to formulate policies which do not take into consideration the main sources of economic change: genetic resources, technological innovation and institutional reform. The policies in practice all over the continent are too static to accommodate the imperatives of rapid economic change. It is important for these countries to initiate research programmes on how to internalize science and technology into the planning process. So far, science and technology are considered as sectors of the academic or educational establishment and exogenous to the process of economic change. As stressed elsewhere, economic systems change as a result of the introduction of new knowledge or information. These changes are non-linear and minor introductions of information may have major effects on the economy. A realization of this fact may go a long way to change planners' attitudes towards the role of technology in economic development.

Conventional approaches to institutional organization, especially those based on Max Weber's notion of rationality, have left Africa with institutional structures that are incapable of responding to scientific and technological imperatives. The institutions tend to force the economy into some form of equilibrium and therefore select out any innovations that might destabilize the system. By doing so, they also effectively suppress all the major sources of innovations and, therefore, of economic change. This rigidity is coupled with the prevalent 'Village Elder' mentality which inhibits the search for new knowledge and challenges to established beliefs. Some African institutions have become analogues of villages or households where the director, acting as an elder, is the ultimate source of ideas and routine; change and flexibility are not usually welcome under such conditions. The system does not reward creativity or innovation; it often rewards mediocrity and conformity since stability is often more preferable to change. Both Weberian rationality and the village elder systems tend to thrive on the fear of uncertainty and novelty.

The future of Africa will depend very much on the degree to which institutional innovations will occur and spread. Alternative economic theories relevant for the continent will have to reflect the diversity that already occurs within it. Conventional theories tend to underrate the significance of diversity and promote uniformity. For development to occur, social systems need a certain measure of diversity, autonomy and the capacity to undertake experiments.[2] It appears that most of the established institutions in Africa tend to inhibit these three requirements. The only institutions that tend to operate on the basis of these requirements are community groups; the government analogue of such institutional arrangements would be decentralization. Already, some African countries such as Kenya have started to decentralize their planning process. This move appears in the short-run to reduce the power of the state, but in the long-run it is likely to lead to rapid changes, especially as the application of science and technology to development advances. On the whole, Africa is caught in a situation where its academics and thinkers must actively participate in the search for alternative worldviews; the grip of old theories is weakening and the future needs not to be predicted, it has to be invented. This is the context in which Africa finds itself in relation to advances in biotechnology.

Science and technology policy

The battle over genetic resources will not be won at international conferences or through international conventions. The ability to benefit from any of these protracted negotiations will depend largely on the ability of the African countries to strengthen their scientific and technological capacity. One of the prerequisites for this is to have an effective science and technology policy. Biotechnology cannot be discussed outside the broader context of scientific and technological development. Effective science and technology policies should lead to the enhancement of the capacity of the African countries to generate technologies. This is a learning process which does not emerge without investment in resources and time. Moreover, it requires strategies that are carefully worked out and reflect the long-term national and regional development needs.

Most African countries have four broad sectors of the economy in which technologies are generated. These are the formal capital goods and service sectors, the so-called informal sector and the scientific and research and academic institutions. The formal sectors are often divided into foreign, and locally-controlled enterprises and joint ventures. Most of the foreign enterprises rely on imported designs of products and processes from their mother firms. The local firms on the other hand import the machinery. In some countries, the capacity to design and produce products and processes exists but there are no effective policies for promoting the process. In some cases, policies on trade and technology imports inhibit the local generation of technology. the informal sector is highly creative in these countries but receives neither the financial backing nor training that is required to improve innovations.

Highly skilled design capability exists in some of the research and academic institutions, but the capability is not effectively linked to the productive sector. The process of linking the design capability with the productive sector is complex and needs to be studied. This is even more crucial given the traditional gap between the scientific establishment and the private sector. Facilitating the process of technology generation or technological innovation may require the formulation of a wide range of policy measures, such as tax incentives, financial assistance, government procurement, industrial property protection, and the strengthening of the R&D institutes.[3] The effectiveness of such measures will depend on the existing policy regime. As mentioned elsewhere, some of the existing policies are inimical to the process of technology generation. Moreover, innovation is risky because of uncertainties in the market environment and it therefore requires supporting measures from both the government and the private sector.

Apart from the machinery already in place, the African countries will still rely on technology importation for a long time to come. Evidence from various countries indicates that defects in the process of technology acquisition have led to excessive project failure. There is a need to look into the current patterns of technology acquisition to identify the key policy issues that have led to these failures. The diversity among African countries and the institutional variation within countries make it difficult to present general statements about technology acquisition. There are, however, some generic issues which are pertinent.

In the first place, the process of technology acquisition implies a selection process; but selection presupposes the existence of technological options. Most *ex post* studies tend to consider technological acquisition in the context of previously existing options. But there are numerous factors that may make it difficult to select technology from the available range of options.[4] For example, the amount of information available on the range of, and on individual options to the potential buyer is crucial in determining the quality of the decision-making process. Imperfections in the information flow make it difficult for technology importers to make well-informed decisions.

Much of the technology flow to the African countries is linked to long-term financial arrangements. The fact that these countries have used foreign exchange to import technology makes finance a major determining factor in technology choice. Complex inter-relationships have, therefore, emerged between government officials, machinery suppliers, foreign banks and consultancy firms on international technology transfer. An understanding of these inter-relationships may provide African countries with insights into the patterns of technology and finance movement and therefore add to their policy learning process.

Most technology acquired by African countries comes from the advanced countries. Recent studies conducted by the United Nations have stressed the importance of South-South trade in technology. This may be a suitable option, especially in intermediate technologies which may be more suited to the condition of the African countries. Not much is known about the possibilities of technology acquisition from other Third World countries and there are numerous cases where machinery is acquired from other Third World countries and sold to Africa through European firms, often at exorbitant prices.[5] Third, the patterns of negotiation play a crucial role in the subsequent operation of imported technology; it is at the level of negotiation that the competence accumulated by various countries is put to the test. The fragmentation of the African countries and their small markets makes it difficult to strengthen their bargaining power. Collective bargaining would not only take advantage of the perceived large markets, but would also enable the countries to specialize in various aspects of technological advancement.

Research into the patterns of collective negotiation is needed, especially in view of the emphasis on collective self-reliance and co-ordination in recent political statements from the African heads of state.[6] The numerous non-technical obstacles that may hinder collective bargaining need to be identified. The issue of negotiation also relates to the legal aspects of international technology transfer. This issue has recently become controversial as a result of recent changes in patent law, especially in biological innovations. Major changes in international legislation are underway to protect innovations resulting from biotechnology and these changes are raising fundamental questions related to the international exchange of genetic resources and the resulting innovations.

It is difficult to determine the central locus of technology policy formulation in most African countries. The existence of institutes, departments or ministries of science and technology does not necessarily show where the policies are formulated and implemented. The very fact that such institutions have a sectoral slant ignores the intersectoral nature of technological change. Science and technology policy should be an integral part of development strategies. This, of course, assumes that African countries have consistent development strategies. It can, therefore, be argued that the introduction of suitable science and technology policies presupposes the existence of long-term economic strategies. Realistically, such strategies vary depending on the specific historical evolution of the countries. Some countries' economic policies are based on agricultural production while others emphasize industrial output. These variations may lead to differences in focus. Science and technology policies are therefore part of the historical development of specific countries.

In this respect, a study of science and technology policy requires an understanding of the economic history of specific countries or regions. Only through this can the historical factors that affect the current patterns of development be identified. Failure to review the historical developments of these countries may lead to the abstract application of certain policies on a uniform basis without considering the existing diversity. Policies that aim at using science and technology to reduce poverty and enhance equity may be different from most of those in operation today. Such policies will have to take into account the fact that the majority of the poor people have little influence over the science and technology system. Efforts must therefore be made to ensure that the focus of technological innovation is relevant to the needs of the poor and of their participation in the process.

It is important to recognize that there are at least two types of policies: explicit and implicit. Explicit policies are often outlined in documents and are associated with some specific implementing institutions; implicit policies are usually linked with practices which are not specifically designated as technology policies. In many cases, implicit policies (such as pricing structures, tariffs, custom duties and export strategies) may have a more significant impact on technological development than explicit policies. This suggests that the study must look beyond those areas that openly manifest themselves as technology policy issues.

Technological development is largely a learning process; it is the introduction of new forms of socio-economic organization. Not only does this rely on the existing knowledge base, but it also creates demands for the generation of technology-specific knowledge. Technological development should, therefore, add to the cumulative knowledge that helps in the formulation of policies. This, however, is not always the case. It is common to find mistakes that are repeated despite the existence of previous failure experiences. This may suggest that the conscious need to learn from project implementation is not usually part of the procedure of implementing industrial projects. As a result, valuable knowledge that could be used in subsequent project implementation processes is lost. It may also be the case that the knowledge is stored at the level of one firm and is not diffused into others. Learning may therefore be occurring at firm and not at a national level.

Facilitating the learning process may require changes in the manner in which projects are implemented, the patterns of project information dissemination and level of public accountability. In some cases, the degree of secrecy surrounding investment projects makes learning more difficult. Moreover, some of the contractual arrangements entered into in the transfer of technology are inimical to learning. For example, in certain contracts, the technical knowledge pertaining to the project may not be disclosed in less than ten years after commissioning. If the project fails before commissioning, it becomes difficult to analyse the reasons for the failure, especially if they relate to the technical details of the plant.[7] Policy formulation is therefore not merely the setting of rigid guidelines but an adaptive process which involves the accumulation and incorporation of new knowledge into the economic system over time. It is influenced both by past development patterns as well as anticipation of development needs.

From recent studies of technology policies in various countries it appears that the policy formulation regime is mediated through institutional organization. It is thus imperative to examine the structure, behaviour and performance of institutions in the various countries. Even in countries such as the US, where free enterprise is the dominant political and economic philosophy, institutions (especially state organs) have played a major role in facilitating technological development.[8] A detailed study of institutional behaviour will uncover some of the major obstacles to innovation. Such findings would contribute significantly to science and technology policy studies. A review of the work of some of the scientific and technical research centres suggests that some fundamental innovations have not taken root in the economic system because of institutional factors. An examination of the efforts to market charcoal stoves in Kenya, diffuse biogas technology in China or introduce producer gas in the Philippines, suggests that institutional factors are crucial to the process of technological innovation. This should, therefore, be seen as a priority area in the formulation of science and technology policy.

The process of technological development, especially the generation of new technological variants, occurs over time and is associated with changes in institutional inter-relationships. Although generic patterns in the role of institutions may be identified, every technology has its own institutional requirements. This realization alone raises a number of research issues. On the one hand, the structure of most institutions does not allow for the flexibility required during the process of technological innovation, and on the other, a loose network of institutions may not provide the long-term commitment and continuity needed to bring technological innovations to maturity.

Because technological change is purposive and is associated with both hierarchical and non-hierarchical forms of organization, it requires an institutional setting that guides the evolutionary process over a long period. There are several reasons why technological development is closely associated with institutional change. In the first place, innovation is a social process which requires networks for bringing different ideas, individuals or groups of people together. The emergence of such a network implies the necessity for institutional arrangements. Second, since the direction and consequences of any evolutionary path are difficult to determine in advance, institutional arrangements are required to provide a broad stage for identifying, selecting and retaining particular goals and tactical methods.

Since an evolutionary process requires the generation, selection and retention of options, policy formulation has a similar pattern. Technology policy may hence be viewed as guidelines for the appraisal, search, generation, selection and retention of technological options. Since the process of implemention is associated with constant change, policies themselves become enriched by the implementation process because of cumulative learning. This would lead us to the partial conclusion that policy itself could be a result of the process of implementation just as implementation could arise from policy.

The policy implementation process takes place in a holistic environment influenced by technical, economic and political factors. Policy formulation and implementation are articulated through institutional organization. This, however, does not mean that the articulation is restricted to institutions alone; far from it. This articulation is a complex interplay between a wide range of actors. The uncertainty in the consequences of any evolutionary route is associated with differences in expectations, strategies and tactics, which suggests that institutions are not homogeneous and manifest extensive internal differences that may lead to conflicts. This is what gives institutions their political character. It is, therefore, no coincidence that national politics (and its ideological underpinnings) is largely about the generation, selection and retention of development options. Politics is about the governance of human evolutionary processes.

The inter-relationship between the various institutions is closely associated with the flow of information, knowledge and other resources, such as finance. Information flow is important because the process of technological innovation involves learning and numerous adaptations both in the technology and the related organizational aspects. Thus institutional organization is an integral part of the process of innovation and economic change. One of the major problems with some conventional economic theories is that institutions are treated as exogenous to the process of technological and economic change. It is assumed that institutions (especially government bodies) distort economic functions and therefore need to be curtailed. The removal of some institutions may have a negative effect on the process of technical change. Since technical change is at the core of economic growth, such steps would undermine the ability of the economies to industrialize. The issue in not whether institutions are required or not; what is at stake is the nature of institutions and their capacity to anticipate or adapt to changes in the economic system.

Institutional innovations need to be accompanied by legal reform and vice versa. It is common to find major legal reforms which are not accompanied by institutional reorganization to effect its implementation. This is often the case where national governments sign international conventions or introduce legislation as a result of international concerns. Alternatively, new institutions may be established without looking into the legal implications of their operations. Existing laws and institutions may be so contradictory as to retard the pace of technological innovation.

The lack of venture capital in Africa is a major obstacle to technological innovation and numerous inventions have gone undeveloped partly for this reason. African countries may need to consider according venture capital tax-free status so that the resources could be utilized in supporting inventive activity. Additionally, national and regional banks should set aside funds that could be used as venture capital for technological innovation. This cause could be furthered if, for example, private investors set up trust funds for supporting innovation and technology policy research. African consultants could also set up private firms in which profits and royalties would be used for funding research.

One of the key aspects of long-term technological development is to provide science education to the population. This could be done through formal and informal education. Most African countries have not yet come to terms with this requirement and therefore still treat science education as one of the optional subjects available to students. Since providing science education requires more financial support for equipment and other facilities, countries with financial or economic problems are most likely to neglect the subject. It is, however, just such countries that are most in need of science education. Some African countries have already started reforming their educational systems to respond to the scientific and technological needs of economic changes. Kenya, for example, has already introduced an educational system that emphasizes the technical sciences and thus is likely to take advantage of some of the advances in biotechnology that those countries which still emphasize the arts cannot. This, however, will take place only in a climate of policy reform and institutional reorganization. One of the ways of ensuring that the population is generally informed in biotechnology is to make the study of biology or related subjects compulsory at all levels of the educational system. Given the long-term implications of modifying life forms, it has become imperative that the public be informed on this vital technology that could easily alter the course of human evolution.

The fostering of a scientific culture has become a key theme for some African scientists and policy makers. Recent attempts to provide a base for the promotion of scientific literacy on the continent include the formation of the Nairobi-based African Academy of Sciences (AAS), which works closely with the Organization of African Unity (OAU) and the Third World Academy (TWA) at Trieste, Italy. The AAS represents a growing awareness among African scientists that the future of the continent depends largely on its ability to integrate science and technology into the culture. This is indeed a major challenge because the African culture is largely based on notions of conformity with the natural environment while the scientific and technological tradition relies largely on intervention and control. The tendency to conform to established norms is also reflected in structure of social organization and local knowledge in traditional societies. A scientific culture leads to questioning and challenging established traditions, a feature that goes contrary to dependence on the elders as the main source of authority and knowledge.

There is an urgent need for African countries to re-examine their culture in view of the imperatives of scientific and technological advances. Some of the features of social organization in Africa, such as decentralized community life, may lead to alternative technologies that are amenable to popular control. The challenge is to establish a viable mode of integration that leads neither to the erosion of the useful features of social organization nor inhibits the application of science and technology to social development. The trend has so far been to replace traditional practices with modern science, often at high social costs. Africa seems to require an alternative view of science; a set of practices that are not based on the expansionist view of Bacon, reductionist approaches of Descartes and the static notions of Newton. Institutions such as the AAS should take up the challenge and identify alternative scientific practices that are consistent with the long-term requirements of the continents.

Changing Africa through science and technology will require major shifts in policy towards the rural sectors. These sectors, in so far as they still remain agricultural, will continue to rely on women as the main sources of economic productivity. The introduction of new technologies will therefore require the involvement of women, starting with the stage of technology design to operation. Only by allowing women to influence the design of agricultural technologies can productivity be improved without major dislocations in the economic system. African countries can take advantages of advances in micro-electronics and produce small-scale technologies which are adapted to the prevailing land tenure and geomorphological conditions.

One of the ways of ensuring that scientific practices reflect the needs of people is to establish close links between the scientific community and local groups. Currently such links are still very minimal and the two groups tend to view each other with suspicion. While most scientists still represent the Cartesian-Newtonian worldview, community groups have tended to distrust the academic establishment. Indeed, many of the problems that community groups are attempting to solve were caused by the same forces that are allied with the scientific establishment. There are, however, numerous scientists who have given much thought to local problems but have not had the opportunity to work with the people. A closer alliance between scientists and community groups will enable the scientific community to define their research in the light of social needs, and community groups will also be able to undertake their work in a more enlightened climate. Moreoever, the alliance will enable the public to work more closely with the scientific community and be able to embody scientific practices into their cultures.

Biotechnology, as pointed out earlier is one those few families of techniques that are amenable to popular participation. They can be applied to decentralized production systems and thus render themselves amenable to local control. Moreover, the public can influence the direction of R&D provided the organization of design and development allows them to do so. It is important that these changes will not occur without deliberate policy efforts. The experience of the Green Revolution has shown how easily the public can lose control of a technology that is meant to serve them. Situations have arisen where increased food production has occurred through the dispossession of land from poor farmers who cannot afford the cheap food produced through the Green Revolution techniques.

Genetic resources and biotechnology

The need for African countries to formulate specific policies on genetic resources is reflected by the limitations of current legislation and the new initiatives to find alternative ways of protecting the resources. One of the first steps would be to undertake a detailed review of existing legislation to identify areas which need strengthening in view of the current conservation imperatives. As pointed out earlier, the legislation covers genetic resources in areas which are designated as 'protected' or where the resources are relevant to the conservation of areas such as wildlife reserves.

This protection does not extend to material that falls outside these regions. Previously, the conservation of wild plant species was undertaken on the basis of their perceived economic or ecological benefits. Wild relatives of species that are currently used tended to be given greater attention. With the increased capacity to screen plants for a wide range of uses, the scope for potentially useful plants is expanding, and national legislation should reflect this change. In addition, the use of biotechnology techniques such as tissue culture and genetic engineering is making conservation possible in areas where it was previously difficult to do so.

Given the growing significance of genetic resources both locally and internationally, it is necessary to look into the possibilities of formulating genetic resources (conservation and utilization) legislation that would harmonize the existing laws, provide administrative measures for conservation, encourage scientific research and facilitate co-ordination between the various institutions involved in genetic resource-based activities. Legislation should also provide for the formation of institutions that would provide technical expertise to other sectors of the economy which utilize genetic resources. There are strong reasons to introduce legislation that would guarantee farmers' rights over local genetic resources as an incentive to conservation. This concept is still underdeveloped, although there are sections of the existing legislation that could be interpreted to show that farmers have rudimentary rights over the material they use. But since the criteria for PBRs are based on 'scientific breeding', those farmers who, over the centuries, have been breeding and selecting the material have no legal claim to it.

Such an arrangement, if linked to the scientific establishment, would also ensure that the local communities benefit from modern breeding techniques without losing control and access over their material. Some seed bank officials claim that they are custodians of the material they hold, but they normally refer to the scientific or research community for whom they claim to be holding the material. National gene banks, if effectively linked into a network of community-based activities, would also serve as custodians of the material for the local farmers. Genetic resource conservation activities should be viewed as a social process and the programme should contribute to the understanding of the way in which the introduction of new material is linked with social reorganization. In this respect, the programmes would, therefore, be organized to include social science studies such as economics, political science, sociology, anthropology and history. Some advances made in the use of genetic resources in medicine have resulted from the collection of ethnobotanical information. A gene bank system should therefore collect local knowledge pertaining to the use and conservation of genetic resources.[9]

Conservation of genetic resources needs to be considered in a regional context. Already, the Nordic Gene Bank has entered into a long-term arrangement with the Southern African Development Coordination Conference (SADCC) countries on germplasm conservation – an agreement that will bring these countries into closer collaboration. Such arrangements need to be undertaken, especially to help meet regional development objectives. Various African countries would be enabled to take advantage of their genetic diversity and specialize in designated crops under such an arrangement.

There seems to be a belief that there are only two ways of conserving genetic resources: *in situ* and *ex situ*. This dichotomy is partly as a result of our limited understanding of the co-evolutionary relationships between genetic resources and socio-cultural change. It is often assumed, especially in the scientific field, that local communities cannot effectively conserve genetic resources, except for minor crops. The focus therefore has been to put excessive emphasis on *ex situ* conservation. This trend has been so widespread in the last 20 years as to attract major criticism,[10] some of which has over-emphasized the opposite view. African countries need genetic resource conservation policies that would build on local knowledge (whenever it is available) while at the same time using modern conservation techniques. The national gene banks could, therefore, be a central body that co-ordinates a network of genetic resource activities involving other research stations, sub-stations, farmers, individual collections and institutional collections. The genetic resources programme could also utilize alternative networks which would include community organizations.

The current efforts to strengthen agricultural research in various African countries offer opportunities to formulate national programmes on genetic resources that would go beyond the currently disjointed and inadequate activities. Such efforts would first of all involve major studies of the current institutions and policies to establish their potential and limitations. In addition, there is a need to establish national capability in the area of genetic resources conservation, both in terms of skilled personnel and equipment. The example of community-based tree planting has shown clearly that a local project, if given policy support and legitimacy, could go a long way in supplementing government initiatives. Like reforestation, the programme would be building on local knowledge and resources. Furthermore, the programme should allow for experimentation and flexibility so that local communities can establish their own methods of conservation within their labour and material limitations.

The role of community-based groups or non-governmental organizations in genetic resource conservation is beginning to be recognized both in the industrialized and Third World countries. The Seed Savers

Exchange (SSE) in the US, for example, has created a network of individuals and groups with seeds that are not available from commercial sources, and in the last decade has provided seed samples of heirloom or unusual garden varieties for nearly 300,000 plantings of non-conventional vegetables not to be found in any seed catalogue; some of the crops were close to extinction.[11] Another US example of grassroots conservation is the North American Fruit Explorers (NAFEX) which has a membership of 3,000 growers who locate, test and preserve superior or special fruit and nut varieties irrespective of their commercial value.

Farmers' organizations in countries such as India, Kenya, Zimbabwe, Tanzania and the Philippines are increasing their involvement in seed exchange and conservation. Other organizations, such as church groups, are also contributing to genetic resource conservation. In Tanzania, for example, a local church at Peramiho has established a sunnhemp (*Crotalaria ochroleuca*) seed bank and has been distributing seed to farmers. Sunnhemp increases soil fertility and organic matter; it also reduces weeds. Currently it is being sold to farmers in large quantities, but the government has been slow in promoting its application, although policy makers have recognized its potential in saving foreign exchange they would otherwise have spent on importing fertilizer.

Seed exchange networks based on home-gardens are prevalent in various parts of Africa. These networks are dominated by women and some of these activities are already becoming so complex that some local women are now growing traditional crops to sell seed. A recent study of Bungonia in Western Kenya has shown that such community networks maintain both exchange and commercial interest in local seed production.[12] So far, the role of community-based groups and networks in genetic resource conservation is not adequately understood and their potential role in maintaining biological diversity is underrated.

In addition to these measures, there are ways of guaranteeing *in situ* conservation in the rural areas by using some of the existing institutions. One possibility is for every school in the country to set up a botanical garden in which local plants would be conserved. This could be linked to the existing school curriculum and the material could be used for teaching purposes. Not only would this ensure the conservation of the local materials, but it would also help to educate the students about the value of local plants. Schools in an administrative location could specialize in the conservation of particular plants to reduce unnecessary duplication. The schools could be linked to other conservation centres. Similar activities could be undertaken by churches, most of which are already involved in the production of vegetables and other crops. Other institutions that could become centres of genetic resource conservation include police stations, district and provincial headquarters as well as local chief's offices. The fact that educational, administrative and religious institutions are available in most parts of Africa could ensure that a large share of the genetic heritage is conserved.

Biotechnologies are being considered as a potential source of increased productivity in African agriculture and forestry, and are also expected to play a significant role in health care. But concerns are already emerging over their potential impacts. The anticipated effects of biotechnology on agriculture are largely based on the experiences of the Green Revolution. Biotechnology, however, is likely to have more profound effects than those of the Green Revolution. Unlike the Green Revolution, the emerging techniques are more suitable to diverse and decentralized production. When issues of equity are considered, however, what matters is not only technical characteristics, but also forms of social organization. Some of the emerging biotechnology techniques, such as tissue culture propagation, are not only likely to have social effects, but they will also reorganize the structure of the scientific community.

It is not known how these technologies will affect the social structure, despite their potential contribution to productivity. Policy research into emerging issues will have two main aspects: first, it is necessary to anticipate the potential contribution of certain technologies and plan the supply of relevant skills in those fields; second, it is important to formulate long-term policies which would shape the development and application of these technologies and the related forms of social reorganization.

Innovation in biotechnology is closely linked to biological diversity; this is the resource base from which the raw material for various technologies is drawn. The conservation of genetic diversity has now moved from the previous ecological concerns and become a basis for technological development. Policies on the development of biotechnology must therefore take into consideration the various techniques for conserving genetic resources. This issue has acquired renewed interest following the recent famines and food insecurity in Africa and the widespread deforestation and environmental degradation. Advances in biotechnology will require increased emphasis on the utilization and conservation of genetic resources. Issues pertaining to intellectual property and legal status of *in situ* and *ex situ* germplasm are increasingly becoming major national and international policy questions. What was originally treated at the FAO meetings as genetic resource questions are now becoming more general biotechnology policy issues.

In addition to these issues, the introduction of new crops into agriculture and industry is likely to change the productive sector of the African countries. The genetic resources collected in the last decade will lead to new products and processes with unpredictable impacts on these countries and the need for them to formulate anticipatory policies will become increasingly imperative. So far, most of the policy issues are based on the potential negative impacts. There is, however, a need to accompany the prognosis with a review of some of the potential benefits of the emerging technologies. It is possible for African countries to establish nascent biotechnology programmes building on areas where research, scientific and technical capability exist.

So far there are no inventories of the current biotechnology-related work in most African countries or the capacity that exists in Africa. Clearly, however, in countries such as Kenya, the existing capacity in the biological and biochemical sciences is sufficient to enable the formulation of a core biotechnology programme. The first step would to be to undertake a survey of current activities, establish the required manpower (as well as institutional arrangements) and pinpoint those areas in which the country has competitive advantage. The next step would be to identify the country's priorities depending on current and anticipated needs as well as the available capability. It is through the identification of priorities that the government can formulate long-term strategies for training as well as for the formation of institutions that would help to meet the programme objectives. This is one of the most difficult aspects of any policy formulation process because priorities tend to change with time and in some cases the dominant scientific traditions may distort the identification of priorities.

Priorities should be identified through open and popular consultation. Efforts should be undertaken to involve groups that do not normally express their views on scientific matters. Sociologists have shown that the identification of research priorities is usually influenced by the micro-politics of the research community and the dominant figures in particular disciplines. Care should, therefore, be exercised to ensure that the process is as representative of the real conditions as possible. This is usually a difficult task because the existing institutions have a tendency to gear their research efforts towards products for which large markets exist; high-income groups often define the research agenda. But the main problems facing African countries relate to the provision of basic needs to communities with low levels of income. As was pointed out earlier, the current international trends in biotechnology are likely to discriminate further against the poor. Changing the trends will require definite policy reorientation by the government.

The requirements of biotechnology research are long-term in nature and will therefore need the commitment of the government over the long-run. The policy commitment would need to transcend the provisions of the short-term development plans and should be organized so that they can survive short-term political changes and policy shifts. Most African countries lack the institutional history to carry forward projects over the long-run; a study of institutional organization in countries such as Japan may be of great help. This would be followed by experimental projects that would be aimed at identifying the most suitable institutional arrangement for the implementation of biotechnology projects.[13] This issue is crucial because every technology has its specific institutional requirements; it cannot be assumed the old ways of organizing institutions will work under new technological conditions. Technological projects in African countries as well as in industrialized countries tend to fail largely because of limitations in institutional organization. The experimental project would require close collaboration between the scientific community, the private sector, banking houses, policy makers and sections of the international community.

In addition to specific experimental projects, there is an urgent need to establish the capacity to monitor international trends so as to assess the potential impacts of the emerging innovations on the economy. The capacity of the country to adapt to the changing conditions may also be found in other developments occurring in the international scene. In this respect, the countries should move away from the 'wait-and-see' policy and participate actively in the global search for biotechnology information and knowledge. Some of the changes occurring in the international scene will have long-term effects, therefore early knowledge of the developments is a prerequisite for timely policy responses.

Most African countries have embassies, high commissions, trade missions and consulates in the industrialized countries. These missions could serve their countries much better if they served as centres for the acquisition of scientific and technological intelligence. This emphasis needs a redefinition of national security priorities. Many of these missions are concerned with trade and military relations, yet the acquisition of scientific and technological capability would have long-term benefits for their countries. This shift in focus will require African countries to locate competent scientists and technologists in their embassies to monitor technological trends and send home any relevant information for the long-term development goals of the continent. The choice of diplomatic ties should be based largely on the long-term national security priorities of the country. At this point in socio-cultural evolution, science and technology are matters of national security. Those countries which ignore the dictates of this reality stand a strong chance of being reduced to mere geographical expressions without any major capacity to transform their economies.

The development of technological capability in biotechnology should be a long-term government objective. As noted earlier, the prospects for the international transfer of biotechnology are diminishing as a result of the restructuring of the research environment in the industrialized countries. Furthermore, the identification of national priorities may require the generation of specific skills needed for the research. This will, therefore, require the country to commit itself to building local capability in the relevant areas. In the short-run, capability-building may be facilitated by the redeployment or retraining of the existing manpower. One option for formal education planning would be to introduce science subjects at much earlier stages with emphasis on the biological sciences.

To supplement these policy requirements, it will be necessary to study the existing laws in order to understand the various ways in which they may hinder or promote the development of biotechnology. Some of the existing laws and the related institutional arrangements may be obstacles to rapid industrial transformation. There is a need to review the laws and keep them in line with the technological imperatives of rapid industrial change. We have also shown how changes in the international legal regime are likely to affect African countries; it is equally important to review the legislation in view of these developments.

If the African countries do not take the initiative in the field of biotechnology as a priority in long-term national security, their scientists are likely to be drawn into research programmes that reflect the agenda of the industrialized countries. Countries such as the US have outlined their interests in maintaining biological diversity in the Third World countries. The reasons for doing so relate largely to the need to protect their political, economic and scientific interests. There are several proposals which will enable the US to use its international influence, especially development assistance, to support programmes that maintain biological diversity in the Third World countries.[14] These programmes will not necessarily be based on the interests of the Third World countries.

The industrialized countries are currently looking into various ways of maintaining global genetic diversity for their national interests. One of the possible ways of gaining access to genetic materials is to enter into arrangements in which sections of the Third World territory will be conserved for scientific or conservation purposes. These possibilities are opened up by the example of Bolivia where, in 1987, some 4.0 million acres of a remote northern part were set aside for conservation under strict legal protection, with a fund of US$250,000 set up to manage the area. This deal was started when the US-based Conservation International, a non-profit organization, with the help of Citicorp Investment Bank, bought up US$650,000 of Bolivia's foreign debt obligations (worth US$4.0 billion), for which an assortment of lenders accepted a total of only US$100,000.

The region now remains accessible not only for conservation but also for the sustainable utilization of its genetic material. The Bolivian example illustrates the numerous ways in which the debt crisis could be a source of long-term economic renewal, but not necessarily for the debtor countries. The challenge is to think of innovative ways in which similar arrangements could contribute not only to the conservation of genetic diversity but also to the accumulation of scientific and technological capability. The amount of money paid by the Conservation International is so small that it could have been contributed by Bolivian nationals, some of whom have their savings in foreign currency.

Agricultural conversion

One of the most significant policy implications of the current developments in biotechnology is the potential for the partial decoupling of the world commodity markets. Over the years there have been those who have argued for closer economic integration between the economies of the Third World countries and those of the industrialized countries through a more equitable commodity trade system, as exemplified by the work of the Brandt. There are those, especially from the left, who have argued for the delinking of the African economies from those of the industrialized countries.[15] Both schools of thought have not adequately taken into consideration the impact of technological innovations on the international commodity market.

What seems to be happening is the partial delinking of the various raw material supply lines and end-use markets as a result of technological innovations. This shift away from dependence on raw materials from the Third World countries is posing new challenges to governments. Most of the emphasis in Third World countries has been devoted to negotiating for better commodity prices. The current developments suggest that some of the countries will have to make decisions which entail shifting away from the supply of raw material to other economic activities. Major adaptations will have to be made in the agricultural sectors of those countries affected by innovations in biotechnology.

The capacity of a country to adjust to major changes in industrial or agricultural production depends on several external and internal factors. In the last 20 years, the global market for raw materials has become increasingly unstable with market niches shrinking and expanding depending on a wide range of factors. One of the main reasons has been the introduction of substitutes for various raw materials. It was predicted that some of these substitutes would replace raw materials from the Third World countries. What has happened, especially in minerals, has been a reduction of the share of some resources as various market niches become occupied by substitutes.[16]

The fact that the substitutes have not made any wholesale replacement of the major raw materials has created a false sense of hope that the traditional resources can make a comeback and command high prices on the international market. Those who advocate the revival of high commodity prices have failed to understand the logic of technological change and direction as well as the rate of change of major innovation.

In the agricultural sector, for example, the new biotechnology products are qualitatively different from the substitutes of the 1960s and 1970s. Producing pyrethrin through tissue culture techniques is qualitatively different from introducing petroleum-based insecticides. Not only are new processes being used, but new natural substitutes are also being used.

One of the responses to the introduction of substitutes was to form trading cartels and argue for commodity stabilization programmes. It is notable that the introduction of substitutes was made largely by firms with different technological traditions from those involved in the buying and selling of dominant commodities. The trends in biotechnology are radically different. The search for new products is funded mainly by the major food companies or importers of the raw materials. Commodity cartels or price stabilization programmes will not effectively work against the long-term strategies of major importers, especially when they have access to technological possibilities for exploring alternative ways to produce the same products.

The introduction of new biotechnology products will therefore be an irreversible process and no major advances are likely to be made at negotiation tables. Under such circumstances, the African countries will have to diversify their policy options. In addition to pursuing international negotiations and possibly reaching some trading agreements with the importing countries in the short-run, efforts should be focused on ways of converting some of the land now under export commodities to other end uses. Such an agricultural conversion programme would have to be anticipatory in nature because it would deal with situations that are yet to be realized.

One of the preconditions for this is to establish an effective monitoring system that would provide adequate technological intelligence on the rate and direction of innovation in biotechnology. This should be a collective effort among the major producers of certain commodities. Cocoa producers, for example, could set up a monitoring unit that would provide up-to-date information on technological trends. The monitoring activities should be linked into national and regional research programmes aimed at providing data on alternative uses for the land currently under commodity production. Where such options include alternative agricultural commodities, the programme would be linked into national or regional agricultural research stations.

This work could be co-ordinated under an agricultural conversion programme involving global monitoring and R&D. But for an effective conversion programme to be implemented, the countries must have reliable access to a wide range of genetic resources from which they can develop optional crops. This means that the future viability of such a conversion programme will depend on current efforts to conserve genetic resources. As noted earlier, if current developments in intellectual property protection in the industrialized countries persist, it will become increasingly difficult to have free access to either genetic resources or improved varieties conserved or developed in the industrialized countries. In this respect, an agricultural conversion programme must go hand-in-hand with efforts to conserve genetic resources.

The various methods of genetic resource conservation have been discussed elsewhere in this study. It was pointed out that one way to ensure that genetic resources are not lost is to entrust their conservation into the hands of farmers. The farmers would not only become custodians of different types of genetic resources, but they can also be incorporated into breeding programmes that emphasize *in situ* conservation. So far most modern agricultural approaches in Africa discourage farmers from growing their indigenous varieties; as a result the genetic resource base is gradually being undermined. On the whole, farmers need incentives to conserve the wide range of genetic diversity available to them for purposes of long-term national agricultural conversion and diversification programmes. Giving these incentives requires the provision of certain rights over the material conserved by the farmers.

The emerging trends in biotechnology may also force the African countries to re-examine some of the dominant policies, especially those aimed at attracting foreign investment in the agricultural sector. There is already a decline in general direct foreign investment to the Third World countries and it is expected that there will be even less being invested in those sectors likely to be affected by biotechnology research. There is a need to shift investment policies from reliance on international investment to the development of local entrepreneurship. These countries need to look into alternative ways of setting up or supporting venture capital activities so that local finance capital and innovation can be more closely linked.

Institutional innovation

In order the respond to the changes in biotechnology and genetic resource conservation, it is necessary to introduce innovative institutional reforms that will facilitate the process of technological change and guarantee the long-term conservation of the available resources. One possibility is to provide farmers with legal rights over their genetic material. One of the reasons for granting farmers' rights is to provide a legal framework within which they can be compensated for their germplasm conservation efforts. The effectiveness of the international fund for plant genetic resources can be enhanced through the existence of farmers' rights over the material they select or domesticate. Under the existing arrangements, the fund could allocate finances to national governments but there are no effective ways whereby the money could be directed to the farmers. There is a large institutional gap between international bodies and farmers; governments have so far failed to establish closer links between the two.

Although national and international institutions are often organized and co-ordinated in their dealings with farmers, the latter are often treated as isolated entities with a very limited organizational basis. Local or grassroot institutions within which the farmers articulate their concerns and views are often ignored or sidestepped in policy formulation. It is, therefore, necessary that farmers' rights be recognized and enshrined in national legislation. The sources of such rights are already embodied in many African traditional laws pertaining to plant tenure. What is needed is a consolidation of these bundles of rights into a coherent legal framework.

Before examining the issue of farmers' rights in detail, it is important to establish the basic philosophical principles that would govern the mode of compensation to farmers for their contribution to the global genetic resource pool. One way of looking at the issue is to apply the 'equity principle' under which the value of germplasm emanating from the Third World countries should compensate for the economic costs of the PBRs system. The system would require complex computations on the flows and counterflows of germplasm of different degrees of improvement across numerous national borders. Countries such as Ethiopia, with large Vavilov centres, are expected to benefit more from an equity-based compensation system.

There are numerous obstacles to the effective functioning of such a system. In the first place, the logistics of tracking down the global flow of germplasm would make the process of compensation extremely complex. Second, the fact that the contribution of specific genes to agriculture is relevant only when viewed in a wider context of crop production makes such a compensation system even more complex. Although the equity principle seems a fair way of compensating Third World farmers, it is too complex to function at an international level.

An alternative to this is the 'incentive principle' under which finances would be made available to encourage farmers to conserve genetic resources and be rewarded for their efforts. One of the ways for implementing such a system is to create conditions under which some rights would be granted to farmers to enable them to bargain for a price for their varieties. This would be similar to the way industrialists use patent licences to establish prices for their innovations. Such a system would work effectively in situations where farmers have monopoly control over certain varieties. But for farmers to benefit from such a system, they would possibly have to form the equivalent of seed companies.

Alternatively, funds would be made available, possibly from a share of the global seed sales (through an international fund) to adequately compensate for the efforts of farmers in genetic resource conservation. But for such an incentive system to work, farmers would require certain rights over their material and these rights would be different from those granted to breeders. Whichever system of compensation is chosen, the need for farmers' rights still remains a significant way to provide a legal framework within which conservation incentives could be provided. The effectiveness of introducing such rights will depend largely on the nature of existing legal systems and their provisions in other biotechnology-related areas. The existing patent systems in sub-Saharan Africa are colonial legacies. Most Francophone countries belong to the Organisation Africaine de la Proprieté Intellectualle (OAPI) based in Yaoundé, Cameroon; they have a uniform patent law. Most of the Anglophone countries belong to the Harare-based African Regional Intellectual Property Organization (ARIPO), whose member states fall into three categories of patent systems.

The first category comprises countries such as Botswana, Lesotho and Swaziland which confer automatic protection to patents registered in South Africa. The second category includes countries which require that patents be granted in the UK before they are re-registered in their countries. These include Gambia, Ghana, Kenya, Seychelles, Sierra Leone, Tanzania and Uganda. This latter category also includes Liberia, Malawi, Mauritius, Nigeria, Somalia, Sudan, Zambia and Zimbabwe which have independent patent systems based either on the United Kingdom law or the model patent law prepared by the WIPO under the auspices of ARIPO.

Kenya, for example, operates a dependent patent system under which local registration of patents is only for those patents granted in the UK. A person issued with a UK patent could apply for patent registration in Kenya within three years of the granting of the UK patent. The system does not require any substantive disclosure of the information. All that is needed is two certified copies of the specifications including drawings, if any, of the UK patent, and a certification of the UK Patent Office Comptroller-General. This patent registration system has proved unsatisfactory and the Kenya government has been preparing an alternative law.

A 'National Council for Science and Technology Legal and Patents Committee' was set up to: revise the law; draw up guidelines for an effective patent system; harmonize patent, trademark and standards policies; and make specific recommendations for a national patenting policy. After consideration, the committee recommended that the WIPO model law be used as a basis for the new patent law. The proposed law will subject applications to substantive examination. This will be done by government experts and in collaboration with ARIPO under the Harare Protocol on Patents and Industrial Designs adopted in 1982. The

Protocol empowered ARIPO to examine, grant and administer patents and industrial design rights on behalf of its signatories. The Harare Protocol came into force in 1984 and was signed by Botswana, the Gambia, Ghana, Kenya, Malawi, Sudan, Uganda, Zambia and Zimbabwe. Given the limited number of skilled people, the early stages of the implementation of the patent law will be based on formal examination. Substantive examination will be undertaken when the skilled people and resources are available.

In addition to the standard provisions of the rights and obligations of the applicant, the proposed law will include compulsory licensing.[17] This will apply to cases where the patents could be, and are not being worked or where the non-working of the patent is prejudicial to Kenya's export earnings. The proposed law also provides for the use of patented inventions by the government. This will apply to inventions that are deemed to be of vital importance to national security, health and the development of strategic sectors of the economy. This will be invoked by a relevant minister who shall direct that the invention be used by the government or an individual authorized by the government.

In order to finance the operations of the patent office, the law will require the payment of reasonable registration fees and maintenance renewal fees. In addition to these provisions, the proposed law will ensure that contracts between local firms and foreign licensers are properly negotiated to facilitate the genuine acquisition of technology. The proposed law maintains the demarcation between biological and inanimate inventions and intends to exclude 'plant or animal varieties or essentially biological processes for the production of plants and animals, other than microbiological processes and the products of such processes'.[18] This demarcation is already being eroded in the industrialized countries and a large number of biological innovations are receiving patent protection. An increasing number of Third World countries are considering reforming their legislation to bring it in line with international developments.

Many of the innovations being produced by the African countries may not qualify for protection under the patent system. These intermediate innovations can, however, be protected through 'petty patents' or utility certificates, which are based on lower levels of novelty. Such a system was extensively used by Germany and Japan in their efforts to upgrade their innovative levels. Much of the innovative activity in the textile sector was fostered by the issue of utility certificates which protected intermediate innovations that relied on the recombination of local and imported technologies. A utility certificate system, as a supplement to the patent system, is easier to administer and has lower financial and administrative costs. The system offers at least five advantages.

First, it can help protect traditional innovations not covered by the patent system and usually displaced by new technologies. The existence of the utility certificate system would enable local innovators to improve their technologies so as to be protected. Secondly, the system would enable local or traditional innovations to play a more significant role in economic development. Under current conditions, new technologies tend to displace traditional skills and therefore weaken the capacity of the local crafts-people to remain self-employed. Thirdly, using utility certificates would enable the society to elevate its innovative levels. This can be done by giving short periods of protection (say, seven years) within which the inventor is required to improve on the invention or risk losing it to a competitor. Fourthly, innovations that can be protected by utility certificates do not require large research investments and can be afforded by large sections of the society. Finally, issuing utility certificates can act as a source of data on the country's innovative activity as well as experience in technological management.[19]

One of the links between *in situ* genetic resources and *ex situ* collection is the availability of ethnobotanical information on the use of various plants – knowledge is part of the culture of various communities. This knowledge, however, and the plants that are associated with it, is not treated as part of the rights of the farmers. Folk knowledge is taken for granted partly because it is not organized as private property. The trend has often been to appropriate the knowledge and the related plant material and turn them into private property.

It is known that local communities have undertaken plant selection activities for hundreds of years; they should therefore have some property rights over the material. Traditional societies have evolved novel ways of plant tenure, which in most societies predates land tenure.[20] But the material is usually treated as unimproved and therefore not subject to protection. The requirements for intellectual property protection for plants assume that the local communities have not undertaken any steps to improve their varieties or that only modern breeding brings about plants that can be protected. This decision is indeed arbitrary and rests on discriminatory assumptions that modern scientific knowledge is superior to other knowledge systems.

The rights of land tenure that farmers enjoy partly rest on the view that land is necessary as an element of continuity in the existence of individuals or social groups. It is understood that to deny farmers this right may endanger their survival and the productive use of the resource. Seeds, however, tend to be treated as isolated entities which do not enjoy the same recognition pertaining to continuity. This is only the case if a static and reductionist view is applied to genetic material: static in that the seed is conceived as being dormant and therefore not directly manifesting any changes; reductionist in the sense that the seed is isolated from a larger process of socio-cultural evolution.

If the seed is taken as an embodiment of genetic information, what the farmers select, breed and conserve are pieces of genetic information which allow for continuity through time. Just as the breeders are granted rights over the discovery or recombination of genetic information, so do farmers deserve similar rights over the collection, selection, improvement and conservation of the material. The rights should, therefore, be accorded in view of the fact that farmers do select and conserve genetic information and not merely some biological material or foodstuff.

As pointed out above, the introduction of new genetic material in a community reorganizes the socio-economic system. Since the introduction is normally conscious, the farmers do assume certain responsibilities over their actions and assert some control over the materials they introduce. From this understanding we argue that farmers should be accorded rights over the materials they use to transform their local economic systems. Although the law does not accord the farmers any specific rights, some of the administrative practices suggest that farmers do have property rights over their plant material. For example, eviction orders are usually timed to follow harvest, otherwise the government would have to pay compensation for the crops left behind. This approach, however, simply recognizes the plants as property in the sense that they are grown for use and that they are simply 'fixtures' on the land. It does not deal with the rights of the farmers over the genetic information contained therein and its reproductive potential.

Nearly all the local communities have customs which are directly or indirectly aimed at *in situ* genetic resource conservation. Kenya uses customary law where the modern law does not apply. This would suggest that in areas where customary law provides for the communal ownership of plants or genetic resources, such protection should, therefore, be legally recognized. What is more difficult to establish, however, is the extent to which the law would be interpreted to cover the material that has been removed from its customary domicile and is located elsewhere, say in foreign gene banks.

In addition to existing legislation, African countries should use contractual protection of some of their genetic material; this form of protection may be more effective than patents. Such arrangements could be analogues of new approaches being proposed to companies and researchers who

> Should not make accessible the samples of organisms or biological products they have developed, without obtaining a clear, binding, written agreement specifying (a) any permissible restrictions on the use of the material loaned to the recipient, e.g., that it is to be used only for basic research purposes, not for commercialization; (b) notifying the recipient that the organization or person lending the material is

retaining title to the material, parts thereof, progeny, products, etc.; (c) that no one else is to receive a sample of any of the provider's micro-organisms, parts thereof, progeny, products, etc. without prior written permission of the provider.[21]

Provision should also be made which requires the recipient to share the technical information and research results from the material with the provider. These measures should be aimed at facilitating the flow of genetic material and the related know-how in a more equitable way.

Most African countries are signatories of the International Undertaking on Plant Genetic Resources and some are members of the FAO Commission on Plant Genetic Resources (CPGR). The signatories of the CPGR, at their 1987 meeting in Rome, reconized the concept of farmers' rights as being as significant as PBRs, although the former is not enshrined in national legislation. The evolution of the concept of farmers' rights seems to be occurring in a manner that would reduce potential conflicts arising from adherence to PBRs. While PBRs guarantee royalties to breeders, compensation to farmers for their contribution to agriculture is now being reflected in the setting up of the fund for the conservation and utilization of genetic resources under the auspices of FAO. It is proposed that the global network of gene banks holding material collected by the IBPGR be brought under the control of FAO. This would enable the compensation system to operate under conditions that would allow the farmers to be indirectly compensated through some form of taxation on the sale of seeds by industry.

There are several actions that African countries can undertake to support the international legal developments. The first would be to introduce analogous arrangements at the national level. A small tax on seed sales could go to setting up a national network of gene banks and to provide incentives to farmers to conserve their varieties. The second step would be to work out a legal regime to establish farmers' rights at a level that would be comparable to the provisions of PBRs. The genetic resource incentive system would therefore be more complete since it would compensate farmers for their conservation efforts, and breeders for their improvement activities. By granting farmers' rights, African countries would also be asserting property rights over material that is in their territories and would therefore strengthen their capacity to bargain with potential users for other forms of compensation. They, for example, may require that the information and technical knowledge arising from the use of genetic material collected from the continent for breeding be exchanged freely with African scientists as a way of building local capability in related fields. This would be a useful way of using the local resources and would probably be more valuable than financial compensation.

Furthermore, only by establishing farmers' rights can local communities effectively benefit from the efforts currently underway at the FAO to rationalize the global genetic resources system. So far, there are no clear ways through which the farmers can benefit from the Fund. It is assumed that benefits of the fund will accrue to the farmers through national governments. The economic history of Third World countries and failure of the so-called 'trickle-down theory' has demonstrated that this does not always take place. But by formulating suitable farmers' rights, the countries can set up institutional arrangements through which the farmers can benefit from the use of the material arising from their conservation and selection activities. In the final analysis, the efforts should lead to the establishment of community gene banks, most likely under the control of women, who are the main sources of botanical knowledge. This should also ensure that technical knowledge related to germplasm conservation and utilization will be entrusted to the hands of women who currently dominate the agricultural scene in Africa.

The continent has so far been active at various international meetings and initiatives. The challenge, however, is to start working on innovative issues. Given the uncertainty over the potential impacts of biotechnology, African countries should seek to acquire capability in this field and at the same time argue for a code of conduct that would enable humanity to respond in time to any negative effects of biotechnology products. Such initiatives could be pursued either through existing international agencies which have technical expertise or through an international commission on a code of conduct on the manipulation of life forms. This code of conduct should aim at enabling humanity to maintain a certain measure of collective preparedness in the face of technological opportunities that can detrimentally change the course of evolution.

This study began by pointing out that the current trends in the industrialized countries are narrowing the development options open to Third World countries in general and Africa in particular. This is happening when these countries are faced by new challenges and are looking into alternative development options. The challenges are represented by an old Chinese saying: if we do not change our direction, we are likely to end up where we are headed. Changing direction implies that the nation or regions understand their current position and can identify the alternative routes open to them. The ability to do so requires extensive flexibility. But flexibility is not enough unless there is enough technological and scientific capability to translate policy changes into practice. Preoccupation with policy reform without building scientific and technology capability is like worrying which way the ship is going without minding the engine room. Most African countries tend to fall into this category. On the other hand, minding the engine room without knowing which way the ship is going is equally futile.

Notes

1. See Bohm and Peat, 1987.
2. This theme is explored in detail in Rosenberg and Birdzell, 1986.
3. SEE UNCTAD, 1986b, for a detailed account of policies and instruments that could be used to facilitate technological innovation in Third World countries. Selected measures are covered by various UNCTAD case studies. Other case studies in the local accumulation of technological capability are presented in Fransman and King (eds), 1984. African countries are starting to examine the role played by patents in acquiring technology as well as stimulating local technological and scientific innovations.
4. The neo-classical school uses the production function to show the existence of a wide range of options that are available by merely making different combinations of labour and capital. This view has lost most of its ground. See Stewart and James, (eds) 1982, for a collection of studies which challenge this view.
5. Recent UNCTAD studies have also looked into the benefits of acquiring technologies from medium-sized and small firms instead of the traditional suppliers of machinery. This focus may also allow the African to import technologies from other Third World countries. Recent evidence from Kenya shows that technologies imported from small and medium-sized firms are more labour-intensive than those acquired from transnational corporations. See UNCTAD, 1986c.
6. This is a major theme in the Lagos Plan of Action as well as the regional economic arrangements that are operating in various regions of Africa. See also United Nations, 1986a.
7. See Juma, 1986, for details of such obstacles to technological learning. The secrecy issue may also relate to the patterns of investment in various countries. For example, in countries where government equity is a major source of investment capital, projects with large government equity participation become parastatal organizations. Information on such projects becomes official secrets and it is then difficult for researchers to have access to it. Moreover, any criticism of such projects (however constructive) may be construed as anti-governmental. Such situations may create a hostile climate to research.
8. See Nelson (ed.), 1982, for case studies on the subject.
9. 'The conservation and use of crop genetic diversity illustrates the international convergence of interests in raising agricultural productivity in the Third World. The tools of biotechnology are needed to store, evaluate, and manipulate the genes in traditional crop varieties and wild plants needed for plant breeding. Yet much of the diversity itself still resides in farmers' fields, where the crops are adapted to the idiosyncrasies of local rainfall, soils, and cultivation methods', Wolf 1986, pp. 38-9. See Richards, 1985, for a study of the role of local knowledge in agricultural production.

10. See Mooney, 1979 and 1983.

11. OTA, 1986b, p. 35.

12. Juma, 1987c.

13. So far the linkages between universities and the productive sectors in Africa are very poor and there have not been any major efforts to ensure that research activities and industrial production are closely linked to ensure economic development. As pointed out earlier, biotechnology is largely a result of university research. The absence of linkages between industry and university institutions will make the formulation of viable biotechnology programmes even more difficult.

14. OTA, 1987, pp. 285-308.

15. This view is held mainly by advocates of the 'Dependency school'.

16. For an explanation of the concept of market niches, see Juma, 1986 and Clark and Juma, 1987.

17. Provisions for compulsory licensing also exist in the current plant varieties protection law.

18. WIPO, 1978, p.19.

19. Juma, 1987b, pp. 20-21.

20. Weinstock and Vergara, 1987, p. 312. See also Fortmann and Riddell, 1985; and Raintree (ed.), 1987.

21. Bent et al., 1987, pp. 382-3.

Conclusions and Conjectures

Africa finds itself at a critical period in world history. While the continent is searching for alternative development options, the industrialized countries are making reforms and introducing innovations that will narrow the development options open to these countries. This study has argued that the future of African economies will depend largely on their capacity to harness some of the advances in biotechnology and apply them in decentralized economic development strategies. The conservation of genetic resources will be crucial for such future development strategies. This issue, as shown in the various chapters, will become increasingly controversial — especially in matters pertaining to the control and ownership of genetic resources.

But control and ownership will be inadequate unless accompanied by the building of scientific and technological capacity to incorporate the resources into the economic system. The question of genetic resources is not only a matter of economic growth. The capacity to manipulate life has provided humanity with enormous possibilities for transforming the world in unpredictable ways. The issue has become a matter of national or regional security.

Biotechnology offers extensive possibilities to deal with Africa's chronic food and medical problems. These, however, are not the only issues facing the continent. The potential for biotechnology to be used for military and subversive objectives is starting to emerge as an issue of concern. Most African countries today lack the capacity to monitor the impact of modified life forms. As a matter of national security, these countries need to build a certain level of technological and scientific capability that will enable them to detect the 'biological smoking gun'. Furthermore, they should initiate and participate fully in efforts to ensure that humanity is well prepared to deal with innovations which have negative impacts, especially given the potential for biotechnology products to detrimentally alter the course of evolution.

Biotechnology has empowered humanity to create as well as destroy life using genetic information. The spectre of biological warfare now hangs over the human race like a bad dream; unfortunately, the Promethean potential that is now being accumulated in various countries makes it a possibility. Never has mankind lived through such moments of prospect and fear before. Biotechnology and genetic resources have now moved to the agenda of national and regional security. Africa cannot continue to ignore the imperative of socio-economic evolution: innovate or perish.

Appendix: Institutions Conserving Genetic Resources

Country/Acronym & Institution		Major Crops
Afghanistan	Plant Genetic Resources Unit, Darulaman*	Large collection of local barley, chickpea, faba bean, lentil, pea and wheat, and introduced germplasm of maize, mung bean, *Phaseolus* and rice
Argentina	Estación Experimental Regional Agropecuaria, Pergomino	Large collection of barley, cotton, flax, oat, forages, groundnut, maize, *Phaseolus,* wheat, sorghum, soyabean and tomato
Australia	Commonwealth Scientific and Industrial Research Organization, Canberra	Introduced forage grasses and legumes, rice, soyabean, sunflower, and indigenous wild *Glycine* species
	Western Australian Department of Agriculture, South Perth	Large collection of introduced forages and legumes
	Australian Wheat Collection Department of Agriculture, Tamworth	Large collection of *Aegilops* and wheat and smaller collections of barley, rye and Triticale
	Queenland Department of Primary Industries, Brisbane	Cluster bean and safflower
	South Australian Department of Agriculture, Adelaide	Large collection of wild species of forage legumes
	CSIRO Division of Tropical Crops	Large collection of tropical grasses and legumes and Pastures, St Lucia
Austria	Landwirtschaftlich-Chemische Bundesversuch-sanstalt, Linz	Landraces and advanced cultivars of barley, oat and wheat and local landraces of *Phaseolus*
AVRC	Asian Vegetable Research and Development Center, Taiwan*	Large collection of Asiatic landraces, advanced cultivars and breeding lines of *Amaranthus,* black gram, chinese cabbage, other brassicas, mug bean, soyabean, sweet potato and tomato
Bangladesh	Bangladesh Jute Research Institute, Dhaka	Landraces, wild species, advanced cultivars, mutants of species of *Corchorus* and *Hibiscus*

Belgium	Faculté des Sciences Agronomiques de l'Etat à Gembloux	Large number of wild species of *Phaseolus* and *Vigna* and related genera
Brazil	National Genetic Resources Centre, Brasilia	Local collections of brassicas, groundnuts and related wild species, maize, *Phaseolus*, soyabean, and tobacco and introduced germplasm of cowpea, rice, sesame and wheat.
Bulgaria	Institute of Introduction and Plant Resources, Sadovo	Barley, brassicas, *Capsicum*, chickpea, rice, cowpea, cucurbits, faba bean, forage grasses and legumes, groundnut, lentil, maize, oat, onion, pea, *Phaseolus*, potato, soyabean, sunflower, rice, rye, tomato, Triticale and wheat
Canada	Plant Gene Resources of Canada, Ottawa	Large collection of introduced barley, maize, brassicas, millets, oat, soyabean, tomato and wheat
	Agriculture Canada Research Station, Saskatoon	Diverse collection of introduced landraces, breeding lines, advanced cultivars and related wild species of brassicas
CATIE	Centro Agronomico Tropical de Investigacion y Enseñanza, Turrialba	Large collections of indigenous landraces, wild species and advanced cultivars of *Capsicum*, cucurbits, *Phaseolus, Sechium edule, Solanum* and tomato
Chile	Banco de Genes, Instituto de Producción y Sanidad Vegetal, Universidad Austral de Chile, Valdivia	Collection of landraces and wild species of potatoes
China	Chinese Academy of Agricultural Sciences, Beijing*	Collection of indigenous brassicas and related wild species, maize, sorghum, rice and wheat
	Beijing Vegetable Research Centre, Beijing	Range of local vegetable landraces
	Taiwan Seed Service, Taichung, Taiwan	Landraces, breeding lines and advanced cultivars of brassicas and advanced cultivars of *Capsicum*
CIAT	Centro Internacional del Agricultura Tropical, Cali, Colombia*	Large collection of forage grasses and legumes. Large collection of New World landraces, advanced cultivars and related wild species of *Phaseolus coccineus, P. lunatus, P. vulgaris* and smaller collections of cowpea and winged bean
CIMMYT	Centro Internacional del Mejoramiento de Maiz y Trigo, El Batan, Mexico*	Large collection of indigenous landraces of maize from Central and South America. Large collection of cultivars and breeding material of wheat

CIP	Centro Internacional de la Papa, Lima, Peru*	Large collection of indigenous Central and South American landraces and advanced cultivars of potato and related wild species
Colombia	Instituto Colombiana Agropecuaria, Bogota*	Large collection of landraces of *Allium,* barley, *Capsicum,* cucurbits, lupins, maize, *Phaseolus,* rice, sorghum, tuber-bearing solanums, tomato and wheat
Cyprus	Agricultural Research Institute, Nicosia*	Large collection of cereals and legumes
Czechoslovakia	Research Institute of Plant Production, Bratislavska	Collection of barley, clover, lentil, lucerne, *Phaseolus,* soyabean, Triticale and wheat
	Maize Research Institute, Trstinka	Collection of maize germplasm
Ecuador	Instituto Nacional de Investigaciones Agropecuarias, Estación Experimental Santa Catalina	Collection of indigenous landraces of *Amaranthus, Lupinus* and quinoa
Ethiopia	Plant Genetic Resources Centre, Addis Ababa	Large collection of indigenous germplasm of flax, barley, brassicas, castor oil, chickpea, cucurbits, faba bean, lentil, millets, niger-seed, pea, safflower, sesame, sorghum, wheat, and introduced *Capsicum,* maize and *Phaseolus*
France	Office de la Recherche Scientifique et Technique Outre-Mer, Bondy	African collection of millets, sorghum and rice
	INRA Station d'Amélioration des Plantes, Le Rheu	Large collection of landraces and breeding lines of cauliflower
	INRA Station d'Amélioration des Plantes Maraîcheres, Montfavet, Avignon	Large collection of introduced landraces and related wild species of *Capsicum,* eggplant and melon and advanced cultivars and breeding lines of tomato
	CNRS Laboratoire de Génétique et Physiologie du Développement des Plantes, Gif sur Yvette	Large collection of millets, rice and sorghum
German Democratic Republic	Zentralinstitut für Genetik and Kulturpflanzenforschung, Gatersleben	World collection of *Aegilops,* barley, *Capsicum,* faba bean, cucurbits, flax, forage grasses and legumes, lentil, lupin, maize, medicinal plants, millets, oat, pea, poppy, *Phaseolus,* rye, sorghum, soyabean, species, tobacco, tomato, vegetables and wheat

GNPG	German-Netherlands Potato Genebank, Braunschweig	Large collection of introduced germplasm of potato and related wild species
Ghana	Crop Research Institute, Bunso*	Collection of landraces of bambarra groundnut, cowpea, lima bean, okra and other local legumes and vegetables
Greece	Greek Genebank, Thessalonika	Local collection of *Aegilops*, barley, beet, brassicas, chickpea, clover, cotton, faba bean, lentil, pea, *Phaseolus* and wheat
Hungary	Research Centre for Agrobotany, Institute for Plant Production and Qualification, Tapioszele	Collection of landraces and advanced cultivars of barley, beet, brassicas, *Capsicum,* chickpea, oat, clover, cucurbits, faba bean, forage legumes, pea, groundnut, lentil, lupin, millets, *Phaseolus,* poppy, sorghum, soyabean, wheat and related wild species and locally collected material of flax, lettuce, maize, onion, sunflower and tomato
ICARDA	International Centre for Agricultural Research in Dry Areas, Aleppo, Syria*	Large collection of *Aegilops,* barley, chickpea, faba bean, forage grasses and legumes, lentil, oat, pea, Triticale and wheat
ICRISAT	International Crops Research Institute for the Semi-arid Tropics, Hyderabad, India*	Large collection of local and introduced chickpea, groundnut, millets, sorghum and pigeon pea
IITA	International Institute of Tropical Agriculture, Ibadan, Nigeria*	Large collection of African germplasm of African yam bean, bambarra groundnut, cowpea and related wild species, kersting's groundnut, lablab bean and rice
		Large collection of cowpea and related wild species, rice and soyabean
India	Central Rice Research Institute, Cuttack	Large collection of rice and related wild species
Indonesia	National Biological Institute, Bogor*	Seeds of indigenous tropical legumes
	Research Institute for Food Crops, Bogor	Large collection of local landraces and advanced cultivars of rice
	Research Institute for Food Crops, Sukamandi	Collection of local landraces and advanced cultivars of rice and introduced mung bean
Iraq	Plant Genetic Resources Unit, Agricultural Research Centre, Baghdad*	Significant collection of local germplasm of *Aegilops,* barley, chickpea, lentil, okra and wheat
IRRI	International Rice Research Institute, Los Baños, Philippines*	Large collection of Asiatic and African landraces, advanced cultivars, breeding lines and wild species of rice

Israel	Agricultural Research Organization, Bet Dagan	Collection of barley, brassicas, chickpea, cotton, forage legumes, lentil, maize, melon, onion, pea, *Phaseolus,* rice, rye, watermelon and wheat
Italy	Instituto del Germoplasmo, CNR, Bari	Large collection of landraces and cultivars of barley, faba bean, forage grasses and legumes, maize, oat, pea, tomato and wheat
	Instituto di Miglioramento Genetico and Produzione delle Sementi, Torino	Collection of landraces and advanced cultivars of *Capsicum,* eggplant and *Phaseolus*
Ivory Coast	Office de la Recherche Scientifique et Technique d'Outre-Mer, Abidjan*	Large collection of indigenous germplasm of cultivated and related wild species of okra and rice
Japan	Plant Germplasm Institute, Kyoto	Large collection of introduced *Aegilops,* barley, *Capsicum, Phaseolus* and wheat
	National Institute of Agro-biological Resources, Yatabe	Large collection of landraces, advanced cultivars and breeding lines of barley, beet, brassicas, cucurbits, eggplant, flax, forages, maize, millets, mung bean, oat, onion, pea, rice, sorghum, soyabean, sweet potato, tomato and wheat
	Vegetable and Ornamental Crops Research Station, Mie	Large collection of landraces of brassicas and landraces, breeding lines and wild tomato species
	Tohoku University, Sendai	Large collection of brassicas and wild relatives
	Faculty of Agriculture, Kobe University, Kobe	Large collection of introduced landraces of potato and related wild species
	Institute of Agricultural and Biological Sciences, Okayama University, Kurashiki	Large collection of barley and related wild species
	Hokkaido National Agricultural Experiment Station, Shimamatsu, Hokkaido	Large collection of introduced landraces and advanced cultivars of potato and smaller holdings of onion
	Laboratory of Plant Breeding, Mie University	Collection of advanced cultivars and landraces of sweet potato
Kenya	National Agricultural Research Station, Kitale*	Large collection of maize, oat and tropical forage species
Korea, Republic of	Germplasm Management Office, Office of Rural Development, Suweon	Large collection of indigenous landraces of rice and wild species of soyabean and locally collected barley, maize, mung bean, red bean, sorghum and wheat

	Agricultural Experiment Station, Suweon	Large collection of landraces and introduced cultivars of rice
Malawi	Chitedze Agricultural Research Station, Lilongwe	Collection of landraces and wild species of cowpea, groundnut, and other minor legumes, maize, millets, rice and sorghum
Malaysia	Rice Research Centre, Rumbong Lima	Large collection of local landraces of rice
Mexico	Instituto Nacional de Investigaciones Agricolas, Mexico City*	Large collection of local cucurbits, maize *Phaseolus* and tropical legumes and introduced chickpea, lentil, medicago, millets, safflower, sesame, sorghum, soyabean, and temperate grasses
Netherlands	Institute of Horticultural Plant Breeding, Wageningen	Large collection of landraces, advanced cultivars and related wild species of brassicas, *Capsicum,* eggplant, lettuce, melon, pea and *Phaseolus*
	Bejo-Zaden, Noordsarwoude	Collection of landraces and advanced cultivars of onion and brassicas
	Foundation of Agricultural Plant Breeding, Wageningen	Large collection of barley, brassicas, faba bean, maize and wheat
NGB	Nordic Gene Bank, Lund	Collection of local and introduced barley, beet, brassicas, clover, flax, forage grasses, oat, onion, pea, rye and wheat
Nigeria	National Horticultural Research Institute, Ibadan*	Collection of landraces, advanced cultivars and breeding lines of *Amaranthus, Capsicum,* cucurbits, okra, onion and tomato
Pakistan	Pakistan Agricultural Research Council, Islamabad*	Collection of landraces of barley, brassicas, okra, chickpea, forages, maize, onion, pea, rice and wheat
Paraguay	Instituto Agronomico Nacional, Cacupe	Collection of native landraces and introduced forage grasses and legumes and wheat
Peru	Programa de Investigaciónes de Maiz, Universidad Nacional Agraria, La Molina, Lima*	Large collection of indigenous maize
	Programa de Investigación en Hortalizas, Universidad Nacional Agraria, La Molina, Lima	Large collection of indigenous *Capsicum,* tomato, cucurbits
Philippines	Institute of Plant Breeding, Unversity of the Philippines, Los Baños*	Large collection of brassicas, *Capsicum,* cowpea, cucurbits, eggplant, groundnut, mung bean, okra, pigeon pea, soyabean, tomato and winged bean

Poland	Institute for Potato Research, Roszalin	Collection of wild species of potato
	Plant Breeding and Acclimatization Institute, Radzikow	Large collection of barley, faba bean, forages, oat, pea, rye, sorghum, soyabean and wheat
Portugal	Maize Breeding Centre, Braga*	Collection of Mediterranean maize
	Departmento del Genética, Estação Agrómica Nacional, Oeiras	Collection of local lupin, maize, *Phaseolus,* rye and wheat
Solomon Islands	Dodo Creek Research Station, Ministry of Agriculture and Lands, Honiara	Collection of landraces of potato
South Africa	Division of Plant and Seed Control, Pretoria	Large collection of introduced advanced cultivars and landraces of barley, brassicas, cotton, oat, cowpea, cucurbits, forage legumes and grasses, groundnut, lupin, onion, *Phaseolus,* rye, sesame, sorghum, soyabean, sunflower, tomato and wheat
Spain	Banco de Germoplasmo INIA, Finca El Encin, Madrid*	Collection of landraces, advanced cultivars and wild species of barley, chickpea, faba bean, forage legumes, lentil, lupin, melon and related species, oat, pea, *Phaseolus,* rye and wheat
	Escuela T.S. de Ingenieros Agrónomos, Universidad Politécnica, Madrid	Large collection of brassicas
	Centro Regional de Investigacion y Desarrollo Agrario del Ebro, Zaragoza	Large collection of *Capsicum*
Switzerland	Station Fédérale de Recherches Agronomique de Changins, Nyon	Large collection of forage grasses and wheat
Syria	Agricultural Research Centre, Douma*	Significant collection of local landraces of *Aegilops,* barley, chickpea, faba bean, forages, lentil, oat, safflower and wheat
Thailand	Thailand Institute of Scientific and Technological Research, Bangkok*	Collection of Asiatic maize and winged bean
	Rice Division, Department of Agriculture, Bangkok	Large collection of landraces, wild types and advanced cultivars of rice
Turkey	Aegean Regional Agricultural Research Institute, Izmir	Large collection of local landraces of barley, beet, brassicas, *Capsicum,* chickpea, cucurbits, forage legumes, lentil, maize, oat, okra, onion, rye,

		Phaseolus, poppy, sesame, spinach, wheat, sunflower, tobacco and related wild species
Uganda	Uganda Agriculture and Forestry Research Organization, Soroti	Large collection of millets and sorghum
USSR	N.I. Vavilov Institute of Plant Industry, Leningrad	World collection of landraces of all major crop species; includes Vavilov's original collection
United Kingdom	Royal Botanical Gardens, Kew	Collection of forage grasses and legumes
	Plant Breeding Institute, Cambridge	Large collection of landraces and wild species of barley, maize, oat, rye and wheat
	National Vegetable Research Station, Wellesbourne*	Large collection of landraces and advanced cultivars of beet, brassicas, carrot, faba bean, lettuce and radish
	Scottish Crop Research Institute, Pentlandfield	Collection of forage crucifers
	John Innes Institute, Norwich	Large collection of landraces, advanced cultivars and wild species of pea
	Department of Applied Biology, University of Cambridge	Large collection of introduced landraces and wild species of *Phaseolus*
	Welsh Plant Breeding Station, Aberystwyth	Large collection of forage grasses
USA	North Eastern Region Plant Introduction Station, Geneva, New York	Large collection of landraces, advanced cultivars and wild species of brassicas, celery, forage grasses and legumes, onion, pea and pumpkin
	National Seed Storage Laboratory, Fort Collins, Colorado	World collections of barley, beet, brassicas, oat, pea, *Capsicum,* castor, cotton, cowpea, cucurbits, flax, rice, maize, forage legumes and grasses, groundnut, lentil, lettuce, millets, okra, onion, *Phaseolus,* cultivated and wild potatoes, rye, safflower, sesame, sorghum, soyabean, tobacco, tomato and wheat
	Southern Region Plant Introduction Station, Experiment, Georgia	Large collection of landraces, advanced cultivars and wild species of *Capsicum,* annual clovers, cucurbits, and other related *Solanum* species, groundnut, *Leucaena* millets, mung bean, okra and pigeon pea, eggplant
	Department of Horticulture, Purdue University, Indiana	Large collection of landraces, advanced cultivars and breeding lines of lima bean

	USDA Vegetable Production Research Unit, Salinas, CA	Collection of advanced cultivars and landraces of brassicas, chicory and lettuce
	North Central Region Plant Introduction Station, Ames, Iowa	Large collection of advanced cultivars and landraces of *Amaranthus,* brassicas, beet, carrot, cucumber, maize, *Medicago,* millets, radish, spinach, sunflower, sweet clover and tomato, *Lathyrus*
	Western Region Plant Introduction Station, Pullman, Washington	Large collection of *Allium,* brassicas, chickpea, faba bean, forage grasses, lentils, lettuce, lupin, *Phaseolus,* safflower and teff
	USDA Small Grains Collection, Plant Genetics and Germplasm Institute, Beltsville	Large collection of barley, oat, rice, rye, Triticale, wild wheat and related genera
	Southern Soybean Collection USDA-ARS Delta Branch Experiment Station, Stoneville	Large collection of soyabean
	USDA Soybean Laboratory, University of Illinois, Urbana	World collection of soyabean
WARDA	West African Rice Development Association, Monrovia, Liberia*	Large collection of African rice and introduced material from Asia

*Designated IBPGR assistance with storage facilities
Source: International Board for Plant Genetic Resources, Rome

Bibliography

Abernathy, T. et. al., (1983) *Industrial Renaissance: Producing a Competitive Future for America,* Basu Books, New York.

Abraham, M. (1987) *The Agrochemical Connection: Pesticides, Seeds, Biotechnology and TNCs,* Paper presented at the 1987 Dag Hammarskjold Seminar on 'The Socio-economic Impact of New Biotechnologies on Basic Health and Agriculture in the Third World,' Bogève (France) and Geneva, 7–14 March.

Adler, R. (1984) 'Biotechnology as an Intellectual Property', *Science,* Vol. 224, pp. 357–63.

Agrow Seed Company (1983) *A Chronicle of Plant Variety Protection: 1983 Update,* Agrow Seed Company, Kalamazoo, Michigan, USA.

Ah von, J. (1986) 'Agricultural Biotechnology and Environment: A Strategy for the Future', *Ceres,* Vol. 19, No. 2, pp. 36–40.

Alan Shawn Feinstein World Hunger Program (1987) *Projecting the Current Perspective: 1957–2057,* Background Paper 2, Workshop on Beyond Hunger: Africa's Future 1957–2057, Kericho, Kenya, 1–5 June.

Alexander, M. (1985) 'Ecological Consequences: Reducing the Uncertainty', Issues in Science and Technology, Vol. 1, 3, pp. 57–68.

Alvares, C. (1986) 'The Great Gene Robbery', *Illustrated Weekly of India,* 23 March, pp. 1–17.

———— (1987) 'Science, Imperialism and Colonialism', *UNU Work in Progress*, Vol. 10, No. 3, p. 11.

Anderson, E. (1969) *Plants, Man and Life,* University of California Press, Berkeley, USA.

Anderson, W.T. (1987) *To Govern Evolution,* Harcourt Brace Jovanovich, New York, USA.

Anon (1986) 'European Parliament Votes to Adopt FAO Undertaking', *Diversity,* No. 9, p. 24.

Antonsson-Ogle, B. (1984) 'Wild Plant Resources', *Ceres,* Vol. 17, No. 5, pp. 38–40.

Arethoon, W.R. and Birch, J.R. (1986) 'Large-Scale Cell Culture in Biotechnology', *Science,* Vol. 232, pp. 1390–5.

Arnold, M.H. et al. (1986) 'Plant Gene Conservation', *Nature,* Vol. 319, p. 615.

Ayala, F.J. (1987) 'Two Frontiers of Human Biology: What the Sequence won't tell us', *Issues in Science and Technology,* Vol. 3, No. 3, pp. 51–6.

Baker, H. (1978) *Plants and Civilization,* Wadsworth Press, Belmont, California, USA.

Balandrin, M. et al. (1985) 'National Plant Chemicals: Sources of Industrial and Medicinal Materials', *Science,* Vol. 228, pp. 1154–60.

Ball, C. (1930) 'The History of American Wheat Improvement', *Agricultural History,* Vol. 4, April.

Baltimore, D. (1987) 'Genome Sequencing: A Small Science Approach', *Issues in Science and Technology*, Vol. 3, No. 3, pp. 48–50.

Bannister, R.C. (1979) *Social Darwinism: Science and Myth in Anglo-American Thought,* Temple University Press, Philadelphia, USA.

Barkin, D. and Suárez, B. (1986) 'The Transnational Role in Mexico's Seed Industry', *Ceres,* Vol. 19, No. 6, pp. 27–31.

Barel, C.D. et al. (1985) 'Destruction of Fisheries in Africa's Lakes', *Nature*, Vol. 315, pp. 19–20.

Bartels, D. (1986) 'Organisational Hazards in Biotechnology — Towards a New Risk Assessment Program', *Prometheus,* Vol. 4, No. 2, pp. 273–7.

Barton, J.H. (1982) 'The International Breeder's Rights System and Crop Plant Protection', *Science,* Vol. 216, pp. 1071–5.

Barton, K.A. and Brill, W.J. (1983) 'Prospects in Plant Genetic Engineering', *Science,* 219, pp. 671–6.

Bateson, G. (1979) *Mind and Nature: A Necessary Unity,* Fontana Paperbacks, London.

Batie, S.S. and Healey, R.G. (1983) 'The Future of American Agriculture', *Scientific American,* Vol. 248, pp. 45–53.

Bean, W.J. (1908) *The Royal Botanic Gardens, Kew: Historical and Descriptive,* Cassell, London.

Bedigian, D. and Harlan, J.R. (1983) 'Nuba Agriculture and Ethnobotany, with Particular Reference to Sesame and Sorghum', *Economic Botany,* Vol. 37, No. 4, pp. 384–95.

Beier, F. et al. (1985) *Biotechnology and Patent Protection: An International Review,* Organization for Economic Co-operation and Development, Paris.

Belsky, J. et al. (1985) *Biotechnology, Plant Breeding, and Intellectual Property: Social and Ethical Dimensions,* Paper for Conference on Ethical Issues in Trade Secrecy and Patent Control Over Research Findings: A Case Study Approach, Center for the Study of Ethics in the Professions, Illinois Institute of Technology, Chicago, June.

Benbrook, C.M. and Moses, P.B. (1986) 'Engineering Crops to Resist Herbicides', *Technology Review,* Vol. 89, No. 8, pp. 55–79.

Bennett, K.J. and Kline, C.H. (1987) 'Chemicals: An Industry Sheds Its Smokestack Image', *Technology Review,* Vol. 90, No. 5, pp. 37–45.

Bennett, W. and Gurin, J. (1977) 'Science that Frightens Scientists: The Great Debate over DNA', *Atlantic Monthly,* Vol. 239, pp. 43–62.

Bent, S.A. et al. (1987) *Intellectual Property Rights in Biotechnology Worldwide,* Macmillan Publishers, London, UK.

Berg, P. et al. (1973) 'Potential Biohazards of Recombinant DNA Molecules', *Science*, Vol. 188, pp. 991–4.

Berland, J.-P. and Lewontin, R.C. (1986) 'Breeders' Rights and Patenting Life Forms', *Nature,* Vol. 322, pp. 785–8.

——— (1986) 'Political Economy of Hybrid Corn', *Monthly Review,* Vol. 38, pp. 35–47.

Bezdek, R.H. et al. (1987) 'Technological Change: Do Scientists Always Benefit?', *Issues in Science and Technology,* Vol. 4, No. 1, pp. 28–34.

Bingham, E.T. (1983) 'Molecular Genetic Engineering vs Plant Breeding', *Plant Molecular Biology,* Vol. 2, pp. 222–4.

Bliss, F.A. (1984) 'The Application of New Plant Technology to Crop Improvement', *Hort Science,* Vol. 19, No. 1, pp. 43–8.

Blumenthal, D. et al. (1986) 'University-Industry Research Relationships in Biotechnology: Implications for the University', *Science,* Vol. 232, pp. 1361–6.

Bohm, D. and Peat, F.D. (1987) *Science, Order and Creativity,* Bantam Books, New York.

Bordwin, H. (1985) 'The Legal and Political Implication of the International Undertaking on Plant Genetic Resources', *Ecology Law Quarterly,* Vol. 12, No. 4, pp. 1053–69.

Borkin, J. (1978) *The Crime and Punishment of I.G. Farben,* The Free Press, New York.

Borlaug, N.E. (1983) 'Contributions of Conventional Plant Breeding to Food Production', *Science,* Vol. 219, pp. 689–93.

Brewbaker, J. (1979) 'Diseases of Maize in the Wet Lowland Tropics and the Collapse of the Classic Maya Civilization', *Economic Botany,* Vol. 33, pp. 101–118.

Brill, W.J. (1985) 'Safety Concerns and Genetic Engineering in Agriculture', *Science,* Vol. 227, pp. 381–4.

Brockway, L. (1979) *Science and Colonial Expansion: The Role of the British Royal Botanic Gardens,* Academic Press, New York, USA.

Brown, L.R. (1970) *Seeds of Change,* Praeger, New York, USA.

——— (1982) *U.S. and Soviet Agriculture: The Shifting Balance of Power,* Worldwatch Paper No. 51, Worldwatch Institute, Washington, DC, USA.

——— and Wolf, E. (1985) *Reversing Africa's Decline,* Worldwatch Paper No. 65, Worldwatch Institute, Washington, DC, USA.

Brown, W.L. (1983) 'Genetic Diversity and Genetic Vulnerability — An Appraisal', *Economic Botany,* Vol. 37, No. 1, pp. 4–12.

Brush, S. (1980) 'Potato Taxonomies in the Andean Agriculture' in Brokensha, D. et al. (eds), *Indigenous Knowledge Systems and Development,* University Press of America, Washington, DC, USA.

Bull, A. et al. (1982) *Biotechnology: International Perspectives,* Organization for Economic Co-operation and Development, Paris.

Busch, L. and Lacy, W.C. (1983) *Science, Agriculture and the Politics of Research,* Westview Press, Boulder, Colorado, USA.

Buttel, F. et al. (1985) 'From Green Revolution to Biorevolution: Some Observations on the Changing Technological Bases of Economic Transformation in the Third World', *Economic Development and Cultural Change,* Vol. 34, No. 1, pp. 31–55.

Calgene (1985) *Annual Report,* Calgene, Davis, California.

Cameron, H.C. (1952) *Sir Joseph Banks, K.B., P.R.S.: The Autocrat of the Philosphers,* The Butterworth Press, London.
Candolle, A. de (1902) *Origin of Cultivated Plants,* Appleton, New York, USA.
Capra, F. (1982) *The Turning Point: Science, Society and the Rising Culture,* Fontana Paperbacks, London.
CAST (Council for Agricultural Science and Technology) (1984) *Development of New Crops: Needs, Procedures, Strategies, and Options,* Report No. 102, CAST, Ames, Iowa, USA.
——— (1985) *Plant Germplasm Preservation and Utilization in U.S. Agriculture,* Report No. 106, CAST, Ames, Iowa, USA.
Cavalieri, L.F. (1981) *The Double-Edged Helix,* Columbia University Press, New York, USA.
CGIAR (Consultative Group on International Agricultural Research) *1985 Annual Report*, CGIAR, Washington, DC, USA.
Chang, T.T. (1984) 'Conservation of Rice Genetic Resources: Luxury or Necessity?' *Science,* Vol. 224, pp. 251–6.
Christensen, C. and Witucki, L. (1982) 'Food Problems and Emerging Policy Responses in Sub-Saharan Africa', *American Journal of Agricultural Economics,* December, pp. 889–96.
Claffey, B. (1981) 'Patenting Life Forms: Issues Surrounding the Plant Variety Protection Act', *Southern Journal of Agricultural Economics,* Vol. 13, pp. 29–37.
Clark, N. (1985) *The Political Economy of Science and Technology,* Blackwell, Oxford, UK.
——— and Juma, C. (1987) *Long-Run Economics: An Evolutionary Approach to Economic Growth,* Pinter Publishers, London, UK.
Clawson, D. (1985) 'Harvest Security and Intraspecific Diversity in Traditional Tropical Agriculture', *Economic Botany,* Vol. 39, pp. 56–67.
Cleaver, H.M., Jnr. (1975) *Origins of the Green Revolution,* PhD dissertation (unpublished), Stanford University, California, USA.
Coats, A. (1969) *The Quest for Plants: A History of the Horticultural Explorers,* Studio Vista, London.
Colwell, R. et al. (1985) 'Genetic Engineering in Agriculture', *Science,* Vol. 229, pp. 111–12.
Connor, S (1986) 'Genes Defend Plant Breeding', *New Scientist,* 27 November, pp. 33–5.
Coggins, G.C. and Harris, A.F. (1987) 'The Greening of American Law? The Recent Evolution of Federal Law for Preserving Floral Diversity', *Natural Resources Journal,* Vol. 27, No. 2, pp. 247 – 307.
Cox, E. (1945) *Plant-Hunting in China: A History of Botanical Explorations in China and Tibetan Marches,* Collins Press, London.
Crabb, A.R. (1947) *The Hybrid-Corn Makers: Prophets of Plenty,* Rutgers University Press, New Brunswick, New Jersey, USA.
Crespi, R. (1982) *Patenting in the Biological Sciences: A Practical Guide for Research Scientists in Biotechnology and the Pharmaceutical and Agrochemical Industries,* John Wiley, Chichester, UK.
Crosby, A.W. Jr. (1972) *The Columbian Exchange; Biological and Cultural*

Consequences of 1492, Greenwood Press, Westport, Conn., USA.
——— (1986) *Ecological Imperialism: The Biological Expansion of Europe, 900–1900,* Cambridge University Press, Cambridge, UK.
Cross, M. (1985) 'Boring into Africa's Grain', *New Scientist,* Vol. 1456, pp. 10–11.
CSTA (Canadian Seed Trade Association) (1984) *Seeds for a Hungry World: The Role and Rights of Modern Plant Breeders*, CSTA, Ottawa.
Culliton, B. (1986) 'Omnibus Health Bill: Vaccines, Drug Exports, Physician Peer Review', *Science,* Vol. 234, pp. 1313–14.
Cunningham, I. (1987) 'Ethiopian Germplasm Repatriated by U.S.', *Diversity,* No. 10, p. 13.
Curtin, P.D. (1964) *Image of Africa,* University of Wisconsin Press, Madison, USA.

Dahlberg, K. (1983) 'Plant Germplasm Conservation: Emerging Problems and Issues', *Mazingira,* Vol. 7, No. 1, pp. 14–25.
——— (1979) *Beyond the Green Revolution: The Ecoloygy and Politics of Global Agriculture,* Plenum, New York, USA.
——— (ed) (1985) *New Directions in Agriculture and Agricultural Research,* Rowman and Allanheld, Totowa, New Jersey, USA.
Dalrymple, D.G. (1986) *Development and Spread of High-Yielding Wheat Varieties in Developing Countries,* US Agency for International Development (USAID), Washington, DC.
——— (1986) *Development and Spread of High-Yielding Rice Varieties in Developing Countries,* USAID, Washington, DC.
Davis, B.D. (1987) 'Bacterial Domestication: Underlying Assumption', *Science,* Vol. 235, pp. 1329, 1332–35.
Day, P.R. (1986) 'The Impact of Biotechnology on Agricultural Research', *Diversity*, No. 9, pp. 33–7.
——— (1987) 'Releasing Engineered Plants: Old and New Concerns', *Swiss Biotech*. Vol. 5, No. 2a, pp. 39–42.
Deardorff, A.V. and Stern, R.M. (1985) *Methods of Measurement of Non-Tariff Barriers,* UN Conference on Trade and Development, Geneva.
de Janvry, A. (1978) 'Social Structure and Biased Technical Change in Agriculture', in Binswanger, H. and Ruttan, V., (eds) *Induced Innovation: Technology, Institutions and Development,* Johns Hopkins University Press, Baltimore, USA.
Dianzungu dia Biniankunu (1987) *Personal Communication,* Brazzaville, Congo.
Dibner, M. (1986) 'Biotechnology in Europe', *Science,* Vol. 232, pp. 1367–72.
Dorozynski, A. (1984) 'The New Technologies: Agricultural Applications', *Ceres,* Vol. 17, No. 6, pp. 15–22.
Doyle, J. (1985) *Altered Harvest; Agriculture, Genetics, and the Fate of the World's Food Supply,* Penguin Books, Harmondsworth, UK.
Dolica, K. (1984) *Understanding DNA and Gene Cloning: A Guide for the Curious,* John Wiley and Sons, New York.
Duvick, D.N. (1957) 'The Use of Cytoplasmic Male Sterility in Hybrid Seed Production', *Economic Botany,* Vol. 13, No. 3, pp. 167–95.

——— (1984) 'Genetic Diversity in Major Farm Crops on the Farm and in Reserve', *Economic Botany,* Vol. 38, pp. 161–78.

ECA (Economic Commission for Africa) (1983) *ECA and Africa's Development 1983–2000: A Preliminary Perspective Study,* ECA, Addis Ababa, Ethiopia.

The Economist (1986) 'Biotechnology's Hype and Hubris', 19 April, pp. 92–3.

——— (1987) 'Fruit Machines', 15 August, pp. 56, 58.

Ehrenfeld, D. (1987) 'Beyond the Farming Crisis', *Technology Review,* Vol. 90, No. 5, pp. 47–56.

Elkington, J. (1986) *Double Dividends? US Biotechnology and Third World Development,* World Resources Institute, Washington, DC.

Elton, C.S. (1958) *The Ecology of Invasions by Animals and Plants,* Chapman and Hall, London.

Erlichman, J. (1987) 'ICI Tightens Grip on World Seeds Supply', *Guardian,* 13 June.

Esquinas-Alcazar, J. (1987) 'Plant Genetic Resources: A Base for Food Security', *Ceres,* Vol. 20, No. 4, pp. 39–45.

Evans, L.T. (1984) 'Darwin's Use of the Analogy Between Artificial and Natural Selection', *Journal of the History of Biology,* Vol. 17, No. 1, pp. 113–40.

FAO (Food and Agriculture Organization) (1981a) *Report of the FAO/UNEP/IBPGR International Conference on Crop Genetic Resources,* FAO, Rome, Italy.

——— (1981b) *Animal Genetic Resources: Conservation and Management,* FAO, Rome, Italy.

——— (1986a) *Legal Status of Base and Active Collections of Plant Genetic Resources,* Commission of Plant Genetic Resources, FAO, Rome, Italy.

——— (1986b) *African Agriculture: The Next 25 Years,* FAO, Rome, Italy.

——— (1987) *Animal Genetic Resources: Strategies for Improved Use and Conservation,* FAO, Rome, Italy.

Faulkner, W. (1986) *Linkages Between Industrial and Agricultural Research: The Case of Biotechnological Research in the Pharmaceutical Industry*, unpublished D.Phil Thesis, Science Policy Research Unit, University of Sussex, Brighton, UK.

Fay, P. (1985) *The Opium War, 1840–1842*, University of North Carolina Press, Chapel Hill, USA.

Fernandes, E.C. et al. (1984) 'The Chagga Homegardens: A Multistoried Agroforestry Cropping System on Mt. Kilimanjaro (Northern Tanzania)', *Agroforestry Systems*, Vol. 2, pp. 73–86.

Fogleman, V.M. (1987) 'Regulating Science: An Evaluation of the Regulation of Biotechnology Research', *Environmental Law*, Vol. 17, No. 2, pp. 183–273.

Fortmann, L. and Riddell, J. (1985) *Trees and Tenure: An Annotated Bibliography for Agroforesters and Others*, Land Tenure Center, University of Wisconsin, Madison, USA.

Fox, J.L. (1981) 'Patents Encroaching on Research Freedom', *Science*, Vol. 224, pp. 1080–82.
——— (1987) 'Contemplating the Human Genome', *BioScience*, Vol. 37, No. 7, pp. 457–60.
Fowle III, J.R. (ed.) (1987) *Application of Biotechnology: Environmental and Policy Issues*, Westview Press, Boulder, Colorado, USA.
Fraley, R.T. (1983) 'Molecular Biology vs Plant Breeding?', *Plant Molecular Biology*, Vol. 2, pp. 49–50.
Frankel, O.H. (1970) 'Genetic Conservation in Perspective', in Frankel, O.H. and Bennett, E. (eds) *Genetic Resources in Plants – Their Exploration and Conservation*, Blackwell Scientific Publications, Oxford, UK.
——— (1970) 'Genetic Conservation of Plants Useful to Man', *Biological Conservation*, Vol. 2, No. 3, pp. 162–9.
——— (1974) 'Genetic Conservation: Our Evolutionary Responsibility', *Genetics*, Vol. 78, pp. 53–65.
——— (1978) 'Conservation of Crop Genetic Resources and their Wild Relatives: An Overview', in Hawkes, J.G. (ed.) *Conservation and Agriculture*, Duckworth, London.
——— (1987) 'Genetic Resources: The Founding Years', *Diversity*, No. 11, pp. 25–7.
Frankel, O.H. and Hawkes, J.G. (eds) (1975) *Crop Genetic Resources for Today and Tomorrow*, Cambridge University Press, London, UK.
Fransman, M. and King, K. (eds) (1984) *Technological Capability in the Third World*, Macmillan Press, London.
Freeman, C. (ed.) (1983) *Long Waves in the World Economy*, Butterworths, London.
Gage, A.T. (1938) *History of the Linnean Society*, Linnean Society, London.
Giamalti, A.B. (1982) 'The University, Industry, and Cooperative Research', *Science*, Vol. 218, pp. 1278–80.
Gilbert, W. (1987) 'Genome Sequencing: Creating a New Biology for the Twenty-First Century', *Issues in Science and Technology*, Vol. 3, No. 3, pp. 26–35.
Godden, D. (1982) 'Plant Variety Rights in Australia: Some Economic Issues', *Review of Marketing and Agricultural Economics*, Vol. 50, No. 1, pp. 51–95.
——— (1984) 'Plant Breeders' Rights and Agricultural Research', *Food Policy*, Vol. 9, No. 3, pp. 206–18.
——— (1987) 'Plant Variety Rights: Framework for Evaluating Recent Research and Continuing Issues', *Journal of Rural Studies*, Vol. 3, No. 3, pp. 255–72.
Goodman, D. et al. (1987) *From Farming to Biotechnology*, Basil Blackwell, London.
Green, H.P. (1983) 'The Impact of Biotechnology on Corporate Law and Regulation', *Technology in Society*, Vol. 5, pp. 87–94.
Griffin, K. (1974) *The Political Economy of Agrarian Change: An Essay on the Green Revolution*, Macmillan Press, London.
Grigg, D.B. (1974) *The Agricultural Systems of the World: An Evolutionary Approach*, Cambridge University Press, Cambridge, UK.

Halluin, A.P. (1982) 'Patenting the Results of Genetic Engineering Research: An Overview', in Plant D.W. et al. (eds) *Patenting of Life Forms*, Banbury Report 10, Cold Spring Harbor Laboratory, Cold Spring, USA.

Halvorson, H.O. et al. (eds) *Engineered Organisms in the Environment: Scientific Issues*, American Association for Microbiology, Washington, DC, USA.

Hansen, M. (1986) 'Plant Breeding and Biotechnology: New Technologies Raise Important Social Questions', *BioScience*, Vol. 36, No. 1, pp. 29–39.

Hanson, J. et al. (1984) *Institutes Conserving Crop Germplasm: The IBPGR Global Network of Genebanks*, International Board for Plant Genetic Resources, Rome, Italy.

Harlan, J.R. (1971) 'Agricultural Origins: Centres and Noncentres', *Science*, Vol. 174, pp. 468–74.

——— (1975a) *Crops and Man*, American Society of Agronomy, Madison, Wisconsin, USA.

——— (1975b) 'Our Vanishing Genetic Resources', *Science*, Vol. 188, pp. 618–21.

Hawkes, J.G. (1983) *The Diversity of Crop Plants*, Cambridge University Press, Cambridge, UK.

——— (1985) *Genetic Resources: The Impact of the International Agricultural Research Centres*, World Bank, Washington, DC.

Hawkins, F. (1944) 'Production of Seeds of Temperate Vegetables in East Africa: War-time Efforts and Suggestions for the Future', *East African Agricultural Journal*, April.

Hayes, H.K. (1957) 'A Half-century of Crop Breeding Research', *Agronomy Journal*, Vol. 49, No. 12, pp. 626–31.

Hayter, E.W. (1968) *The Troubled Farmer, 1850–1900*, Northern Illinois University Press, De Kalb, Illinois, USA.

Heiser, C.B. Jnr. (1973) *Seed to Civilization: The Story of Man's Food*, Freeman, San Francisco, USA.

——— (1979) 'Economic Botany: Past and Future', *Economic Botany*, Vol. 40, No. 3, pp. 261–6.

——— (1979) 'Origins of Some Cultivated New World Plants', *Annual Review of Ecology and Systematics*, Vol. 10, pp. 309–26.

Hiraoka, L.S. (1987) 'Frontiers of Commercial Biotechnology: US and Japanese Potential in a New Industry', *Futures*, Vol. 19, No. 5, pp. 528–44.

Ho, P. (1955) 'The Introduction of American Food Plants into China', *American Anthropologist*, Vol. 57, pp. 191–201.

Hobbelink, H. (1987) *New Hope or False Promise? Biotechnology and Third World Agriculture*, International Coalition for Development Action, Brussels, Belgium.

Hobhouse, H. (1985) *Seeds of Change: Five Plants that Transformed Mankind*, Sidgwick and Jackson, London.

Holden, J.H. and Williams, J.T. (eds) (1984) *Crop Genetic Resources: Conservation and Evaluation*, George Allen and Unwin, Boston, USA.

Holm, L.G. et al. (1977) *The World's Worst Weeds: Distribution and Biology*, University of Hawaii, Honolulu.

Hood, L. and Smith, L. (1987) 'Genome Sequencing: How to Proceed',

Issues in Science and Technology, Vol. 3, No. 3, pp. 36–46.
Hutt, P.B. (1983) 'University/Corporate Research Agreements', *Technology in Society*, Vol. 5, pp. 107–18.
Hymowitz, T. (1984) 'Dorsett-Morse Soybean Collection Trip to East Asia: 50 Year Retrospect', *Economic Botany*, Vol. 38, No. 4, pp. 378–88.

IBPGR (International Board for Plant Genetic Resources) (1980) *Report of a Meeting of a Working Group on Coffee*, IBPGR, Rome, Italy.
——— (1983) *Plant Varieties Rights and Genetic Resources*, IBPGR, Rome, Italy.
——— (1984a) *Collection of Crop Germplasm: The First Ten Years, 1974–1984*, IBPGR, Rome, Italy.
——— (1984b) *The ECP/GR: An Introduction to the European Co-operative Programme for Conservation and Exchange of Crop Genetic Resources*, IBPGR, Rome, Italy.
——— (1987) *Progress on the Development of the Register of Genebanks*. IBPGR, Rome, Italy.
IDRC (1986) *Biotechnology: Opportunities and Constraints*, International Development Research Centre, Ottawa, Canada.
IGRP (1985) 'The Origins of the Genetic Supply Industry', *IDOC Internazionale*, Vol. 17, No. 2, pp. 28–31.
IAA (Institute for Alternative Agriculture) (1986) *Biotechnology and Agriculture: Implications for Sustainability*, IAA, Greenbelt, Maryland, USA.
Ingersoll, F. (1983) 'Private Financing of Public Research', *Seed World*, February, pp. 19–20.
Innes, N.L. (1982) 'Patents and Plant Breeding', *Nature*, Vol. 298, p. 786.
Irons, E.S. and Sears, M.H. (1975) 'Patents in Relation to Microbiology', *Annual Review of Microbiology*, Vol. 29, pp. 319–32.
ITC (International Trade Centre) (1987) *Cocoa: A Trader's Guide*, ITC, UNCTAD/GATT, Geneva, Switzerland.

Jain, H.K. (1982) 'Plant Breeders Rights and Genetic Resources', *Indian Journal of Genetics*, Vol. 42, No. 2, pp. 121–8.
Janzen, D.H. (1973) 'Tropical Agroecosystems', *Science*, Vol. 182, pp. 1212–19.
Jefferson, Thomas, quoted in Brockway, L. (1979) *Science and Colonial Expansion: The Role of the British Royal Botanic Gardens*, Academic Press, New York, p. 35.
Jennings, F. (1975) *The Invasion of America: Indians, Colonialism, and the Cant of Conquest*, University of Carolina Press, Chapel Hill, USA.
Jianming, J. (1987) 'Protecting Biological Resources to Sustain Human Progress', *Ambio*, Vol. 16, No. 5, pp. 262–6.
Joravsky, D. (1970) *The Lysenko Affair*, University of Chicago Press, Chicago, USA.
Joyce, C. (1986) 'US Exports Genetic Experiments', *New Scientist*, No. 1535, 20 November.
——— (1987a) 'The Race to Map the Human Genome', *New Scientist*, Vol. 113, No. 1550, pp. 35–40.

——— (1987b) 'Genes Reach the Medical Market', *New Scientist*, Vol. 115, No. 1569, pp. 45–51.
Judson, H.F. (1979) *The Eighth Day of Creation: Makers of the Revolution in Biology*, Simon and Schuster, New York, USA.
Juma, C. (1985) 'Market Restructuring and Technology Acquisition', *Development and Change*, Vol. 16, No. 1, pp. 39–59.
——— (1986) *Evolutionary Technological Change: The Case of Fuel Ethanol in Developing Countries*, unpublished D.Phil. Thesis, Science Policy Research Unit, University of Sussex, Brighton, UK.
——— (1987a) *Emerging Biotechnologies in Developing Countries: Technological Innovation and Institutional Change*, Paper presented at the 1987 Dag Hammarskjold Seminar on 'The Socio-economic Impact of New Biotechnologies on Basic Health and Agriculture in the Third World', Bogeve (France) and Geneva (Switzerland), 7–14 March.
——— (1987b) 'Patents and Appropriate Technology: The Case for "Utility Certificates"', *Appropriate Technology*, Vol. 14, no. 3 pp. 20–21.
——— (1987c) *Ecological Complexity and Agricultural Innovation: The Use of Indigenous Genetic Resources in Bungoma, Kenya*, Meeting on 'Farmers and Agricultural Research: Complementary Methods', Institute of Development Studies, University of Sussex, UK, 27–31 July.
——— (1987d) *Towards a More Relevant Patent System for Africa: Rapporteur's Report*, Roundtable Meeting, 21–22 January, African Academy of Sciences, Nairobi.
——— (1988) *Rust in Peace: Technological Failure and Development Policy in Africa* (in preparation).

Kalton, R.R. and Richardson, P. (1983) 'Private Sector Breeding Programs: A Major Thrust in US Agriculture', *Diversity*, No. 5, pp. 16–18.
Kanani, S. (1983) *The First Report of the National Task Force on the Biological Control of* Salvinia molesta *on Lake Naivasha*, National Task Force for the Biological Control of *Salvinia molesta*, National Environment and Human Settlements Secretariat, Ministry of Environment and Natural Resources, Nairobi, Kenya.
Kaplinsky, R. (1982) *Computer-Aided Design*, Frances Pinter, London.
Keeler, K.H. (1985) 'Implications of Weed Genetics and Ecology for the Deliberate Release of Genetically-engineered Crop Plants', *Recombinant DNA Technical Bulletin*, Vol. 8, No. 4, pp. 165–72.
KENGO (Kenya Energy Non-Governmental Organizations Association) (1986) *Seeds and Genetic Resources in Kenya*, KENGO, Nairobi, Kenya.
Kennedy, D.M. (1987) 'What's New at the Zoo?' *Technology Review*, Vol. 90, No. 3, pp. 67–73.
Kenney, M. (1987) *The Impact of the International Political Economy on National Biotechnology Programs*, National Symposium on the Role of Biotechnology in Crop Protection, 23–25 January, Kalyani, West Bengal, India.
——— (1986) *Biotechnology: the University-Industry Complex*, Yale University Press, New Haven, Conn., USA.
——— (1985) 'Biotechnology: Prospects and Dilemmas for Third World

Countries', *Development and Change*, Vol. 16, No. 1, pp. 61–92.

Keya, S.O. et al. (1986) 'MIRCENs: Catalytic Tools in Agricultural Training and Development', *Impact of Science and Technology*, Vol. 142, pp. 142–51.

King, J. (1978) 'New Diseases and New Niches', *Nature*, Vol. 276, pp. 407.

——— (1982) 'Patenting Modified Life Forms: The Case Against', *Environment*, Vol. 24, No. 6, pp. 38–41, 57–8.

King, R. (1976) *The World of Kew*, Macmillan Press, London.

Kingdon, C. and Newcombe, S. (1987) 'Engineers Have Designs on Biology', *New Scientist*, Vol. 114, No. 1558, pp. 58–9.

Kjekshus, K. (1977) *Ecology Control and Economic Development in East African History: The Case of Tanganyika, 1850–1950*, Heinemann Books, London.

Kloppenburg, J., Jnr. (1988) *First the Seed. The Political Economy of Plant Biotechnology: 1492–2006*, Cambridge University Press, Cambridge, UK.

——— and Kleinman, D.L. (1987a) 'Seeds of Struggle: The Geopolitics of Genetic Resources', *Technology Review*, Vol. 90, No. 2, pp. 47–53.

——— and ——— (1987b) 'Seeds and Sovereignty', *Diversity*, No. 10, pp. 29–33.

——— and ——— (1987c) 'The Plant Germplasm Controversy', *Bioscience*, Vol. 37, No. 3, pp. 190–98.

Klose, N. (1950) *America's Crop Heritage: The History of Foreign Plant Introduction by the Federal Government*, Iowa State College Press, Ames, Iowa, USA.

Kolato, G. (1985) 'How Safe are Engineered Organisms?', *Science*, Vol. 224, pp. 34–9.

Krimsky, S. (1984) 'Corporate Academic Ties in Biotechnology: A Report on Research Progress', *GeneWatch*, Vol. 1, No. 5/6, pp. 3–5.

——— (1987a) 'The New Corporate Identity of the American University', *Alternatives*, Vol. 14, No. 2, May/June, pp. 20–29.

——— (1987b) 'Gene Splicing Enters the Environment: The Socio-Historical Context of the Debate Over Deliberate Release', in Fowle III, J.R. (ed.) *Application of Biotechnology: Environmental and Policy Issues*, Westview Press, Boulder, Colorado, USA.

Kumamoto, J. et al. (1987) 'Mystery of the Forbidden Fruit: Historical Epilogue on the Origin of the Grapefruit, *Citrus paradisi* (Rutaceae)', *Economic Botany*, Vol. 4, No. 1, pp. 97–107.

Kux, M. (1986) *Land Use Options to Conserve Biological Resources in Developing Countries*, Mimeo, University of Florida, USA.

Lange, P. (1985) 'The Nature of Plant Breeders' Rights (Plant Variety Protection Law) and their Demarcation from Patentable Inventions', *Plant Variety Protection Gazette and Newsletter*, No. 44, pp. 17–27.

Laszlo, E. (1987) *Evolution: The Grand Synthesis*, New Science Library, Boston, USA.

Laufer, B. (1929) 'The American Plant Migration', *Scientific Monthly*, Vol. 28, pp. 239–51.

Lemmon, K. (1968) *The Golden Age of Plant Hunters*, Phoenix House, London.
Lenski, R.E. and Levin, B.R. (1985) 'Genetic Engineering', *Issues in Science and Technology*, Vol. 1, No. 4, pp. 13–14.
Leonard, D.E. (1974) 'Recent Developments in Ecology and Control of the Gypsy Moth', *Annual Review of Entomology*, Vol. 19, pp. 197–229.
Lesser, W.H. (1987) 'Plant Breeders' Rights and Monopoly Myths', *Nature*, Vol. 330, p. 215.
——— and Masson, R.T. (1983) *An Economic Analysis of the Plant Variety Protection Act*, American Seed Trade Association, Washington, DC, USA.
Levin, S.A. and Harwell, M.A. (1986) 'Potential Consequences of Genetically Engineered Organisms', *Environmental Management*, Vol. 10, No. 4, pp. 495–513.
Levins, R. and Lewontin, R. (1985) *The Dialectical Biologist*, Harvard University Press, Cambridge, USA.
Lewin, R. (1987) 'Politics of Genome', *Science*, Vol. 235, p. 1453.
——— (1987) 'Ecological Invasions Offer Opportunities', *Science*, Vol. 238, pp. 752–3.
Lewis, W.H. and Elvin-Lewis, M.P. (1977) *Medical Botany. Plants Affecting Man's Health*, John Wiley and Sons, London.
Lipp, F. (1976) 'A Heritage Destroyed: The Lost Gardens of Ancient Mexico', *Garden Journal*, Vol. 26, pp. 184–8.
Lipton, M. and Longhurst, R. (1985) *Modern Varieties, International Agricultural Research, and the Poor*, World Bank, Washington, DC, USA.
Lower, R.L. (1984) 'Genetic Engineering: The Relationship between Industry, Academia and Plant Sciences', *HortScience*, Vol. 19, No. 1, pp. 49–51.
Luckham, M. (1959) 'The Early History of the Kenya Department of Agriculture', *East African Agricultural Journal*, Vol. 25, No. 2.

Mabogunje, A. (1987) *Personal Communications*, Ibadan, Nigeria.
McAuliffe, S. and McAuliffe, K. (1981) *Life for Sale*, Coward, McCann and Geoghegan, New York, USA.
Maher C. (1950) 'Soil Conservation', in Matheson, J.K. and Bovill, E.W. (eds) *East African Agriculture*, Oxford University Press, London.
Mangelsdorf, P.C. (1974) *Corn: Its Origin, Evolution and Improvement*. Harvard University Press, Cambridge, USA.
Marshall, E. (1981) 'The Summer of The Gypsy Moth', *Science*, Vol. 213, pp. 991–3.
Marx, J.L. (1985) 'Plant Gene Transfer Becomes a Fertile Field', *Science*, Vol. 230, pp. 1148–50.
——— (1987) 'Assessing the Risks of Microbial Release', *Science*, Vol. 237, pp. 1413–17.
Marx, K. (1976) *Capital*, Vol. 1, Penguin Books, Harmondsworth, UK.
Mayor, E. (1982) *The Growth of Biological Thought*, Belknap Press, Cambridge, USA.
Medvedev, Z. (1969) *The Rise and Fall of T.D. Lysenko*, Columbia University Press, New York, USA.

Miller, H. (1987) 'The Case of Qualifying "Case by Case"', *Science*, Vol. 236, p. 133.

Mintz, S.W. (1985) *Sweetness and Power: The Place of Sugar in Modern History*, Penguin Books, Harmondsworth, UK.

Miyata, M. (1985) 'Japanese Corporate Biotechnology R & D', *Science and Technology in Japan*, April/June, pp. 15–20.

Monya, N. (1985) 'The Legal Protection of Achievements in Biotechnology, as Seen by a Japanese Lawyer', *Plant Variety Protection Gazette and Newsletter*, No. 44, pp. 36–43.

Moock, J.L. (ed.) (1986) *Understanding Africa's Rural Households and Farming Systems*, Westview Press, Boulder, Colorado, USA.

Mooney, H.A. and Drake, J.A. (eds) (1986) *Ecology of Biological Invasions of North America and Hawaii*, Springer-Verlag, New York, USA.

Mooney, P. (1979) *Seeds of the Earth: A Private or Public Resource?*, Canadian Council for International Co-operation, Ottawa, Canada.

——— (1983) 'The Law of the Seed: Another Development and Plant Genetic Resources', *Development Dialogue*, No. 1–2, pp. 1–173.

——— and Fowler, C. (eds) (1986) *The Community Seed Bank Kit*, Rural Advancement Fund International, Pittsboro, NC, USA.

Morgan, D. (1979) *Merchants of Grain*, Viking Press, New York, USA.

Morgan, T. (1981) 'Ethnobotany of the Turkana: Use of Plants by a Pastoral People and their Livestock in Kenya', *Economic Botany*, Vol. 35, No. 1, pp. 96–130.

Morwood, W. (1974) *Traveller in a Vanished Landscape: The Life and Times of David Douglas*, Readers Union, Newton Abbot, Devon, UK.

Moses, P.B. and Hess, C.E. (1987) 'Getting Biotech into the Field', *Issues in Science and Technology*, Vol. 4, No. 1, pp. 35–41.

Müller, H.-P. (1981) *Karl Marx: Die technologischhistorischen Exzerpte*, Ullstein, Berlin, West Germany.

Munz and Kech (1973) *A California Flora*, University of California Press, USA.

Murray, J. (1986) 'The First $4 Billion is the Hardest', *Bio/Technology*, Vol. 4, pp. 293–6.

Muturi, S. (1986) *Germplasm Conservation and Utilization in Kenya*, Paper for the National Seminar on Germplasm Conservation and Seed Technology, 13 August, Nairobi, Kenya.

Myers, N. (1979) *The Sinking Ark: A New Look at the Problem of Disappearing Species*, Pergamon Press, London.

——— (1983) *A Wealth of Wild Species: Storehouse for Human Welfare*, Westview Press, Boulder, Colorado, USA.

——— (1985) 'Endangered Species and the North-South Dialogue', in Hall, D.O. et al. (eds) *Economics of Ecosystems Management*, Dr W. Junk Publishers, Dordrecht, The Netherlands.

Nabhan, G.P. (1985) *Gathering the Desert*, University of Arizona Press, Tucson, Arizona, USA.

NAS (National Academy of Sciences) (1972) *Genetic Vulnerability in Major Crops*, NAS, Washington, DC, USA.

——— (1981) *The Water Buffalo: New Prospects for an Underutilized*

Animal, NAS, Washington, DC, USA.
——— (1982) *Priorities in Biotechnology Research for International Development*, NAS, Washington, DC, USA.
——— (1983) *Little-Known Animals with a Promising Economic Future*, NAS, Washington, DC, USA.
——— (1985) *Amaranth: Modern Prospects for an Ancient Crop*, NAS, Washington, DC, USA.
Nelkin, D. (1982) 'Intellectual Property: The Control of Scientific Knowledge', *Science*, Vol. 216, pp. 704–8.
Nelson, R. (ed.) (1982) *Government and Technical Change: A Cross Industry Analysis*, Pergamon Press, New York, USA.
——— and Winter S. (1982) *An Evolutionary Theory of Economic Change*, Belknap Press, Cambridge, Mass., USA.
Newmark, P. (1987) 'Plant Genetic Systems Get Basta Resistance', *Bio/Technology*, Vol. 5, p. 321.
Niklasson, S. (1987) *Personal Communication*, Swedish Patent Office, January.
Noble, D.N. (1977) *America by Design*, Oxford University Press, New York, USA.
——— (1984) *Forces of Production*, Alfred A. Knopf, New York, USA.

Odum, E.P. (1985) 'Biotechnology and the Biosphere', *Science*, Vol. 229, p. 1338.
OECD (Organization for Economic Cooperation and Development) (1986) *Recombinant DNA Safey Considerations*, OECD, Paris, France.
Ogura, T. (ed.) (1966) *Agricultural Development in Modern Japan*, Juji Publishing Co., Tokyo, Japan.
Oldfield, M.L. (1984) *The Value of Conserving Genetic Resources*, US Department of the Interior, Washington, DC, USA.
——— and Alcorn J.B. (1987) 'Conservation of Traditional Agroecosystems', *BioScience*, Vol. 37, No. 3, pp. 199–208.
Oliver, F.W. (ed.) (1913) *Makers of British Botany*, Cambridge University Press, Cambridge, UK.
Oster, G. and Oster, S. (1985) 'The Great Breadfruit Scheme', *Natural History*, Vol. 94, pp. 34–41.
OTA (Office of Technology Assessment) (1981) *Impacts of Applied Genetics*, OTA, Congress of the US, Washington, DC, USA.
——— (1984a) *Commercial Biotechnology: An International Analysis*, OTA, Congress of the US, Washington, DC, USA.
——— (1984b) *Africa Tomorrow: Issues in Technology, Agriculture, and US Foreign Aid*, OTA, Congress of the US, Washington, DC, USA.
——— (1985) *Innovative Biological Technologies for Lesser Developed Countries*, OTA, Washington, DC, USA.
——— (1986a) *Technology, Public Policy, and the Changing Structure of American Agriculture*, OTA, Congress of the US, Washington, DC, USA.
——— (1986b) *Grassroots Conservation of Biological Diversity in the United States*, Background Paper No. 1, OTA, Washington, DC, USA.
——— (1986c) *Technology, Public Policy and the Changing Structure of*

American Agriculture, OTA, Congress of the US, Washington, DC, USA.
——— (1987a) *New Developments in Biotechnology – Background Paper: Public Perceptions of Biotechnology*, OTA, Congress of the US, Washington, DC, USA.
——— (1987b) *Technologies to Maintain Biological Diversity*, OTA, Congress of the US, Washington, DC, USA.

Paine, R.T. and Zaret, T.M. (1975) 'Ecological Gambling: The High Risks and Rewards of Species Introductions', *Journal of American Medical Association*, Vol. 231, No. 5, pp. 71–3.
Parbery, D. (1964) 'Plant Introduction in Asia and Australia', *Nature*, Vol. 202, No. 4932, pp. 549–51.
Pearse, A. (1980) *Seeds of Plenty, Seeds of Want: Social and Economic Implications of the Green Revolution*, Clarendon Press, Oxford, UK.
Perdue, R. et al. (1986) '*Vernonia galamensis*, Potential New Crop Source of Epoxy Acid', *Economic Botany*, Vol. 40, No. 1, pp. 54–68.
Perelman, M. (1977) *Farming for Profit in a Hungry World*, Allanheld, Osmun and Co., Montclair, New Jersey, USA.
Perpich, J.G. (1983) 'Genetic Engineering and Related Biotechnologies: Scientific Progress and Public Policy', *Technology in Society*, Vol. 5, pp. 27–47.
Perrin, R.K. et al. (1983) *Some Effects of the U.S. Plant Variety Protection Act of 1970*, Economics Research Report No. 46, Department of Economics and Business, North Carolina State University, Raleigh, NC, USA.
Perron, C. (1984) *Normal Accidents: Living with High Risk Technologies*, Basic Books, New York, USA.
Persons, S. (ed.) (1950) *Evolutionary Thought in America*, Yale University Press, USA.
Peterson, C.E. (1975) 'Plant Introductions in the Improvement of Vegetable Cultivars', *HortScience*, Vol. 10, No. 6, pp. 575–9.
Pimentel, D.L. (1987) 'Down on the Farm: Genetic Engineering Meets Ecology', *Technology Review*, January, pp. 24–30.
Pineiro, M. and Trigo, E. (eds) (1983) *Technical Change and Social Conflict in Agriculture: Latin American Perspectives*, Westview Press, Boulder, Colorado, USA.
Piore, M.J. and Sabel, C.F. (1984) *The Second Industrial Divide: Possibilities for Prosperity*, Basic Books, New York, USA.
Plant, D.W. et al. (eds) (1982) *Patenting of Life Forms*, Banbury Report 10 Cold Spring Harbor Laboratory, USA.
Plant, D.W. (1983) 'The Impact of Biotechnology on Patent Law', *Technology in Society*, Vol. 5, pp. 95–105.
Plucknett, D.L. et al. (1982) 'Agricultural Research and Third World Food Production', *Science*, Vol. 217, pp. 215–20.
——— (1983) 'Germplasm Conservation and Developing Countries', *Science*, Vol. 220, pp. 163–8.
——— (1986) 'Sustaining Agricultural Yields', *BioSciences*, Vol. 36, No. 1, pp. 40–45.
——— (1987) *Gene Banks and the World's Food*, Princeton University

Press, Princeton, New Jersey, USA.
Plucknett, D.L. and Smith, N.J. (1984) 'Networking in International Agricultural Research', *Science*, Vol. 225, pp. 989–93.
——— (1986a) 'International Prospects for Cooperation in Crop Research' in Consortium for International Cooperation in Higher Education, *Solving World Hunger: The U.S. Stake*, Seven Locks Press, Cabin John, USA.
——— (1986b) 'Benefits of International Collaboration in Agricultural Research', *Economic Botany*, Vol. 40, No. 3, pp. 298–309.
Popovsky, M. (1984) *The Vavilov Affair*, Archon Books, Hamden, Conn., USA.
Popper, K. (1959) *The Logic of Scientific Discovery*, Hutchinson, London, UK.
——— (1963) *Conjectures and Refutations*, Routledge and Kegan Paul, London, UK.
——— (1972) *Objective Knowledge: An Evolutionary Approach*, Oxford University Press, London, UK.
Posey, D.A. (1985) 'Indigenous Management of Tropical Forest Ecosystems: The Case of the Kayapo Indians of the Brazilian Amazon', *Agroforestry Systems*, Vol. 3, pp. 139–58.
Prescott-Allen, R. and Prescott-Allen, C. (1982) *What's Wildlife Worth?*, Earthscan, London, UK.
——— (1983) *Genes from the Wild*, Earthscan, London, UK.
Prigogine, I. (1986) 'Science, Civilization and Democracy: Values, Systems, Structures and Affinities', *Futures*, Vol. 18, No. 4, pp. 493–507.
——— and Stengers, I. (1984) *Order Out of Chaos: Man's New Dialogue with Nature*, Flamingo, London, UK.
Pyenson, L. (1982) 'Cultural Imperialism and Exact Sciences: German Expansion Overseas 1900–1930', *History of Science*, Vol. 20, Part 1, No. 47, pp. 1–43.

Rada, J. (1982) 'Technology and the North-South Division of Labour', in Kaplinsky, R. (ed.) 'Comparative Advantage in an Automating World', *IDS Bulletin*, Vol. 13, No. 2.
RAFI (Royal Advancement Fund International) (1986) 'New Substitutes Threaten to Displace Export Market for Water Soluble Gums', *Bio-communique*, September, RAFI.
——— (1986) 'Bovine Growth Hormone' *Bio-communique*, October/November, RAFI.
——— (1987a) 'Vanilla and Biotechnology', *Bio-communique*, January, RAFI.
——— (1987b) 'Biotechnology and Natural Sweeteners: Thaumatin' *RAFI Communique*, February, RAFI.
——— (1987c) 'Animal Patents', *RAFI Communique*, June, RAFI.
——— (1987d) 'A Report on the Security of the World's Major Gene Banks', *RAFI Communique*, July, RAFI.
——— (1987e) 'Herbicide Resistance', *RAFI Communique*, December, RAFI.
Rains, L.J. (1987) 'Ins and Outs of the New Drug Export Law: Intermediates

and Tropical Diseases', *Bio/Technology*, Vol. 5, p. 304.

Raintree, J.B. (ed.) (1987) *Land, Trees and Tenure: Proceedings of an International Workshop on Agroforestry held in Nairobi, May 27–31, 1985*, Land Tenure Center, Madison, Wisconsin, USA.

Rasmussen, W.D. (1955) 'Diplomats and Plant Collectors: The South American Commission, 1817–1818', *Agricultural History*, Vol 29, No. 1, pp. 22–31.

——— (1975) *Agriculture in the United States: A Documentary History*, Vols. 1–4, Random House, New York, USA.

Rathmann, G.B. (1987) 'An Industry View of the Public Policy Issues in the Development of Biotechnology', in Lowe III, J.R. (ed.), *Application of Biotechnology: Environmental and Policy Issues*, Westview Press, Boulder, Colorado, USA.

Reed, C. (ed.) (1976) *Origins of Agriculture*, Mouton Publishers, The Hague, The Netherlands.

Regal, P.J. (1985) 'Genetic Engineering', *Issues in Science and Technology*, Vol. 1, No. 4, pp. 14–15.

Reichert, W. (1980) 'Agriculture's Diminishing Diversity', *Environment*, Vol. 24, No. 9, pp. 6–11, 39–43.

Republic of Kenya (1972), *Laws of Kenya*, The Grass Fires Act, Chapter 327, Government Printer, Nairobi, Kenya.

——— (1977a) *Laws of Kenya*, The Wildlife (Conservation and Management) Act, Chapter 376, Government Printer, Nairobi, Kenya.

——— (1977b) *Laws of Kenya*, The Seed and Plant Varieties Act, Chapter 326, Government Printer, Nairobi, Kenya.

——— (1977c) *Laws of Kenya*, The Crop Production and Livestock Act, Chapter 321, Government Printer, Nairobi, Kenya.

——— (1979) *Laws of Kenya*, The Plant Protection Act, Chapter 324, Government Printer, Nairobi, Kenya.

——— (1980a) *Laws of Kenya*, The Agriculture Act, Chapter 313, Government Printer, Nairobi, Kenya.

——— (1980b) *National Livestock Development Policy, Ministry of Livestock Development*, Government Printer, Nairobi, Kenya.

——— (1981) *National Food Policy*, Government Printer, Nairobi, Kenya.

——— (1982) *Laws of Kenya*, The Forests Act, Chapter 385, Government Printer, Nairobi, Kenya.

——— (1983a) *Laws of Kenya*, The Suppression of Noxious Weeds Act, Chapter 325, Government Printer, Nairobi, Kenya.

——— (1983b) *Development Plan, 1984-1988*, Government Printer, Nairobi, Kenya.

——— (1984) *Laws of Kenya*, The Antiquities and Monuments Act, Chapter 215, Government Printer, Nairobi, Kenya.

——— (1986) *Economic Management for Renewed Growth*, Sessional Paper No. 1 of 1986, Government Printer, Nairobi, Kenya.

Rheenen, H. van (1979) 'Diversity of Food Beans in Kenya', *Economic Botany*, Vol. 33, No. 4, pp. 448–54.

Richards, P. (1985) *Indigenous Agricultural Revolution: Ecology and Food Production in West Africa*, Hutchinson, London, UK.

Rickett, H.W. (1956) 'The Origin and Growth of Botanic Gardens', *The Garden Journal of New York Botanical Garden*, Vol. 6, No. 5, pp. 133–5, 157–9.

Rifkin, J. (1983) *Algeny*, Viking Press, New York, USA.

Roach, F.A. (1985) *Cultivated Fruits of Britain: Their Origin and History,* Basil Blackwell, Oxford, UK.

Roberts, L. (1987a) 'Who Owns the Human Genome?', *Science,* Vol. 237, pp. 358–61.

——— (1987b) 'Agencies Vie Over Human Genome Project', *Science,* Vol. 237, pp. 486–8.

——— (1987c) 'Human Genome: Questions of Cost', *Science,* Vol. 237, pp. 1411–12.

Rosell, S. (1987) *Personal Communication,* International Development Research Centre, Ottawa, Canada.

Rosenberg, N. (1976), *Perspectives on Technology,* Cambridge University Press, Cambridge, UK.

——— (1982) *Inside the Black Box,* Cambridge University Press, Cambridge, UK.

——— and Birdzell, L.E. (1986) *How the West Grew Rich: The Economic Transformation of the Industrial World,* I.B. Tauris, London, UK.

Ross, E. (1946) 'The United States Department of Agriculture During the Commissionership', *Agricultural History,* Vol. 20, July.

Rossiter, M.W. (1975) *The Emergence of Agricultural Science: Justus Leibig and the Americans, 1840–1880,* Yale University Press, New Haven, Conn., USA.

Ruivenkamp, G. (1986) 'The Impact of Biotechnology on International Development: Competition Between Sugar and Sweeteners', *Vierteljahresberichte*, No. 103, pp. 89–101.

Ruttan, V. (1980) 'Bureaucratic Productivity: The Case of Agricultural Research', *Public Choice,* Vol. 35, pp. 529–47.

——— (1982) 'Changing Role of Public and Private Sectors in Agricultural Research', *Science,* Vol. 216, pp. 23–9.

Ryerson, K. (1933) 'History and Significance of the Foreign Plant Introduction Work of the United States Department of Agriculture', *Agricultural History,* Vol. 7, pp. 110–28.

——— (1976) 'Plant Introductions', *Agricultural History,* Vol. 50, No. 1, pp. 110–28.

Saito, H. (1985) 'Biotechnology R. & D.: Japan and the World', *Science and Technology in Japan,* April/June, pp. 8–11.

Saxonhouse, G. (1985) 'Biotechnology in Japan: Industrial Policy and Factor Market Distortions', *Prometheus,* Vol. 3, No. 2, pp. 227–315.

Schmid, A. (1985) 'Biotechnology, Plant Variety Protection, and Changing Property Institutions in Agriculture', *North Central Journal of Agricultural Economics,* Vol. 7, No. 2, pp. 129–38.

Schmitt, H. (1983) 'Biotechnology and the Lawmakers', *Science in Society,* Vol. 5, pp. 5–14.

Schmitz, A. and Seckler, D. (1970) 'Mechanized Agriculture and Social

Welfare: The Case of the Tomato Harvester', *American Journal of Agricultural Economics,* Vol. 52, pp. 569–71.
Schoenermark, H. von (1974) 'Nathaniel Bagshaw Ward: How a Sphinx Moth Altered the Ecology of the Earth', *Garden Journal,* Vol. 24, pp. 148–54.
Schumpeter, J. (1934) *The Theory of Economic Development: An Inquiry into Profits, Capital, Credit, Interest, and the Business Cycle*, Harvard University Press, Cambridge, Mass., USA.
——— (1939) *Business Cycles: A Theoretical, Historical, and Statistical Analysis of the Capitalist Process,* Vol. 1, McGraw-Hill, New York, USA.
——— (1943) *Capitalism, Socialism and Democracy,* George Allen and Unwin, London.
Schweber, S. (1980) 'Darwin and Political Economists: Divergence and Character', *Journal of the History of Biology,* Vol. 13, No. 2, pp. 195–286.
Shand, H. (1987) 'Cacao & Biotechnology: A Report on Work in Progress', *RAFI Communque,* May.
Sharma, R. (1987) 'Micro-nutrient Drain', *Ileia,* Vol. 3, No. 1.
Sharples, F.E. (1983) 'Spread of Organisms with Novel Genotypes: Thoughts from an Ecological Perspective', *Recombinant DNA Technical Bulletin,* Vol. 6, No. 2, pp. 43–56.
——— (1987a) 'Regulations of Products of Biochemistry', *Science,* Vol. 235, pp. 1329–32.
——— (1987b) 'Applications of Introduced Species Models to Biotechnology Assessment' in Lowe III, J.R. (ed.), *Application of Biotechnology: Environmental and Policy Issues,* Westview Press, Boulder, Colorado, USA.
Shetler, S. (1967) *The Komarov Botanical Institute: 250 Years of Russian Research,* Smithsonian Institution Press, Washington, DC, USA.
Simmons, N.W. (ed.) (1976) *Evolution of Crop Plants,* Longman, London, UK.
Singer, M. and Soll, D. (1973) 'Guidelines for DNA Hybrid Molecules', *Science,* Vol. 181, p. 114.
Slatkin, K. (1987) 'Gene Flow and the Geographic Structure of Natural Populations', *Science,* Vol. 236, pp. 787–92.
Smartt, J. (1984) 'Gene Pools in Grain Legumes,' *Economic Botany,* Vol. 38, No. 1, pp. 24–35.
Sondahl, M. et al. (1984) 'Application for Agriculture', in *Tissue Culture Technology and Development,* ATAS Bulletin No. 1, United Nations Centre for Science and Technology for Development, New York, USA.
Sprague, G.F. et al. (1980) 'Plant Breeding and Genetic Engineering: A Perspective', *BioScience,* Vol. 30, No. 1, pp. 17–21.
Stahl, F.W. (1987) 'Genetic Recombination', *Scientific American,* Vol. 256, No. 2, pp. 91–101.
Staub, W.J. and Blase, M.G. (1971) 'Genetic Technology and Agricultural Development', *Science,* Vol. 173, pp. 119–23.
Steele, A. (1964) *Flowers for the King: The Expedition of Ruiz and Pavon and the Flora of Peru,* Duke University Press, Durham, New Carolina, USA.
Stemler, A.B. et al. (1977) 'The Sorghums of Ethiopia', *Economic Botany,* Vol. 31, pp. 446–60.

Stewart, F. and James, J. (eds) (1982) *The Economics of New Technology in Developing Countries,* Frances Pinter, London.

Stotzky, G. and Babich, H. (1984) 'Fate of Genetically-Engineered Microbes in Natural Environments', *Recombinant DNA Technical Bulletin,* Vol. 7, No. 4, pp. 163–88.

Stowe, R. (1987) 'United States Foreign Policy and the Conservation of Natural Resources: The Case of Tropical Deforestation', *Natural Resources Journal,* Vol. 27, No. 1, pp. 55–101.

Summers, G. (ed.) (1983) *Technology and Social Change in Rural Areas,* Westview Press, Boulder, Colorado, USA.

Sun, M. (1986a) 'Fiscal Neglect Breeds Problems for Seed Banks', *Science,* Vol. 231, pp. 329–30.

——— (1986b) 'The Global Fight over Plant Genes', *Science,* Vol. 231, pp. 445–7.

——— (1986c) 'Local Opposition Halts Biotechnology Test', *Science,* Vol. 231, pp. 667–8.

Swaminathan, M. (1982) 'Biotechnology Research and Third World Agriculture', *Science,* Vol. 218, pp. 967–72.

Sylvester, E.J. and Klotz, L.C. (1983) *The Gene Age: Genetic Engineering and the Next Industrial Revolution,* Charles Scribner's Sons, New York, USA.

Szybalski, W. (1985) 'Genetic Engineering in Agriculture', *Science,* Vol. 229, pp. 112–13.

Tak, J.D. (1984) 'Microbiology Patent Information', *World Patent Information,* Vol. 6, No. 1, pp. 4–11.

Tangley, L. (1987) 'Who Owns Human Tissues and Cells?', *BioScience,* Vol. 37, No. 6, pp. 376–8.

Taylor, L.P. (1987) 'Management: Agent of Human Cultural Evolution', *Futures,* Vol. 19, No. 5, pp. 513–27.

Thomas, C. (1985) *Sugar: Threat or Challenge?: An Assessment of the Impact of Technological Developments in the High-Fructose Corn Syrup and Sucrochemicals Industries,* International Development Research Centre, Ottawa, Canada.

Thomas, W.L. (ed.) (1956) *Man's Role in Changing the Face of the Earth,* University of Chicago Press, Chicago, USA.

Torrey, J. (1985) 'The Development of Plant Biotechnology', *American Scientist,* Vol. 73, July-August, pp. 354–63.

Tucker, J. (1986) 'Amaranth: The Once and Future Crop', *BioScience,* Vol. 36, No. 1, pp. 9–13.

Turrill, W.B. (1959) *The Royal Botanic Gardens at Kew: Past and Present,* Herbert Jenkins, London, UK.

UNCTAD (United Nations Conference on Trade and Development) (1972) *Guidelines for the Study of the Transfer of Technology to Developing Countries,* UNCTAD, New York, USA.

——— (1975) *The Role of the Patent System in the Transfer of Technology to Developing Countries,* UNCTAD, Geneva, Switzerland.

——— (1981) *Examination of the Economic, Commercial and Development Aspects of Industrial Property in the Transfer of Technology to Developing Countries: Review of Recent Trends in Patents in Developing Countries*, UNCTAD, Geneva, Switzerland.

——— (1986a) *Impact of New and Emerging Technologies on Trade and Development: A Review of the UNCTAD Secretariat's Research Findings*, UNCTAD, Geneva, Switzerland.

——— (1986b) *Promotion and Encouragement of Technological Innovation: A Selected Review of Policies and Instruments*, UNCTAD, Geneva, Switzerland.

——— (1986c) *Trends in International Transfer of Technology to Developing Countries by Small and Medium-sized Enterprises*, Report by the UNCTAD Secretariat, UNCTAD, Geneva, Switzerland.

UNDP (UN Development Programme) (1986) *Orientation and Areas of Concentration for the Regional Programme for Africa During the Fourth Programming Cycles (1987–1991)*, UNDP, New York, USA.

UNEP (UN Environment Programme) (1980) *Genetic Resources: An Overview*, UNEP, Nairobi, Kenya.

——— (1988) *Report of the Third Meeting*, UNID/WHO/UNEP Working Group on Biotechnology Safety, UNEP, Nairobi, Kenya.

UNIDO (United Nations Industrial Development Organization) (1982) *Elements of Some National Policies for Biotechnologies*, UNIDO, Vienna, Austria.

United Nations (UN) (1963) *Science and Technology for Development: Report of the UN Conference on the Application of Science and Technology for the Benefit of the Less Developed Areas*, Vol. III, UN, New York.

——— (1975) *The Role of the Patent System in the Transfer of Technology to Developing Countries*, UNCTAD, Geneva, Switzerland.

——— (1986a) *The Critical Economic Situation in Africa: Report of the Secretary-General*, Preparatory Committee of Whole for the Special Session of the General Assembly on the Critical Economic Situation in Africa, UN, New York, USA.

——— (1986b) *United Nations Programme of Action for African Economic Recovery 1986–1990*, Preparatory Committee of the Whole for the Special Session of the General Assembly on the Critical Economic Situation in Africa, UN, New York, USA.

——— (1986c) *Working Paper on the Critical Economic Situation in Africa: Note by the Secretariat*, Preparatory Committee of the Whole for the Special Session of the General Assembly on the Critical Economic Situation in Africa, UN, New York, USA.

United Nations University (UNU) (1985) *The Science and Praxis of Complexity*, UNU, Tokyo, Japan.

United States Congress (1981) *The Department of Agriculture Can Minimize the Risk of Potential Crop Failures*, US Congress, CED-81-75, Washington, DC, USA.

Valerie, M.F. (1987) 'Regulating Science: An Evaluation of the Regulation of Biotechnology Research', *Environmental Law*, Vol. 17, No. 2, pp. 183–273.

Vavilov, N.I. (1951) *The Origin, Variation, Immunity and Breeding of Cultivated Plants,* Roland Press, New York, USA.

——— (1971) 'The Problem of the Origin of the World's Agriculture in the Light of the Latest Investigations', in Bukharin, N. et al. (eds), *Science at the Crossroads,* Frank Cass and Co., London, UK.

Vietmeyer, N. (1986) 'Lesser-Known Plants of Potential Use in Agriculture and Forestry', *Science,* Vol. 232, pp. 1379–84.

Walsh, J. (1984) 'Seeds of Dissension Spout at FAO', *Science,* Vol. 227, pp. 147–8.

Watson, J.D. (1968) *The Double Helix,* Atheneum, New York, USA.

WCED (World Commission on Environment and Development) (1987), *Food 2000,* Zed Books, London, UK.

——— (1987), *Our Common Future,* Oxford University Press, Oxford, UK.

Webb, N. (1987) 'Taking American Biotechnology Across the Atlantic', *Bio/Technology,* Vol. 5, pp. 222–9.

Weinstock, J.A. and Vergara, N.T. (1987) 'Land or Plants: Agricultural Tenure in Agroforestry Systems', *Economic Botany,* Vol. 41, No. 2, pp. 312–22.

Wheale, P.R. and McNally, R.M. (1986) 'Patent Trend Analysis: The Case of Microgenetic Engineering', *Futures,* October, pp. 638–57.

Whitaker, T.W. (1979) 'The Breeding of Vegetable Crops: Highlights of the Last Seventy-Five Years', *HortScience,* Vol. 4, No. 2, pp. 84–6.

White, L. (1967) 'The Historical Roots of Our Ecologic Crisis', *Science,* Vol. 155, pp. 1201–7.

Whitehead, A. (1926) *Science and the Modern World,* Cambridge University Press, Cambridge, UK.

Wilkes, H.G. (1972) 'Maize and its Wild Relatives', *Science,* Vol. 177, pp. 1071–77.

——— (1983) 'Current Status of Crop Plant Germplasm', *CRC Critical Reviews on Plant Science,* Vol. 1, No. 2, pp. 133–81.

——— (1984) 'Germplasm Conservation Towards the Year 2000: Potential for New Crops and Enhancements of Present Crops', in Yeatman, C.W. et al. (eds) Plant Genetic Resources: A Conservation Imperative, *AAAS Selected Symposium 87,* American Association for the Advancement of Science, Washington, DC, USA.

Williams, C.G. (1928) 'The Responsibility of the Agricultural Experiment Station in the Present Agricultural Situation', *Science,* Vol. 67, pp. 519–22.

Williams, S.B. (Jnr) (1984) 'Protection of Plant Varieties and Parts as Intellectual Property', *Science,* Vol. 225, pp. 18–23.

——— (1986) 'Utility Product Patent Protection for Plant Varieties', *Trends in Biotechnology,* Vol. 4, No. 2, pp. 33–9.

Wilson, E.O. (1975) *Sociobiology: The New Synthesis,* Harvard University Press, Cambridge, USA.

——— (1985) 'The Biological Diversity Crisis: A Challenge to Science', *Issues in Science and Technology,* Vol. 2, No. 1, pp. 20–29.

Winkelmann, R. (1981) *Karl Marx: Exzerpte uber Arbeiteilung, Maschinerie und Industrie,* Ullstein, Berlin, West Germany.

WIPO (World Intellectual Property Organization) (1978) *Model Law for English-Speaking African Countries on Patents*, WIPO, Geneva, Switzerland.

Withers, L.A. (1987) '*In Vitro* Methods for Collecting Germplasm in the Field', *Plant Genetic Resources Newsletter*, No. 69, p. 206.

Witt, S.C. (1985) *BriefBook: Biotechnology and Genetic Resources*, California Agricultural Lands Project, San Francisco, USA.

——— (1986) 'FAO Still Debating Germplasm Issues', *Diversity*, No. 8, pp. 24–5.

Wolf, E.C. (1986) *Beyond the Green Revolution: New Approaches for Third World Agriculture*, Worldwatch Institute, Washington, DC, USA.

——— (1987) *On the Brink of Extinction: Conserving the Diversity of Life*, Worldwatch Paper No. 78, Worldwatch Institute, Washington, DC, USA.

World Bank (1984) *Towards Sustained Development in Sub-Saharan Africa: A Joint Programme of Action*, World Bank, Washington, DC, USA.

Worster, D. (1985) *Nature's Economy: A History of Ecological Ideas*, Cambridge University Press, Cambridge, UK.

Yamada, S. (1967) 'Changes in Output and in Conventional and Non-conventional Inputs in Japanese Agriculture Since 1880', *Food Research Institute Studies*, Vol. 7, No. 3, pp. 371–413.

Yarwood, C.E. (1970) 'Man-made Plant Diseases', *Science*, Vol. 168, pp. 218–20.

Yeatman, C.W. et al. (1984) *Plant Genetic Resources: A Conservation Imperative*, Westview Press, Boulder, Colorado, USA.

Yoxen, E. (1983) *The Gene Business: Who Should Control Biotechnology?*, Pan Books, London, UK.

Zaret, T.M. and Paine, R.T. (1973) 'Species Introduction in a Tropical Lake', *Science*, Vol. 182, pp. 449–55.

Zedan, H. (1987) *Personal Communication*, United Nations Environment Programme, Nairobi, Kenya.

——— and Olembo, R. (1985) *A Network of Microbiological Resources Centres (MIRCENs) for Environmental Management and Increased Bioproductivity in Developing Countries*, United Nations Environment Programme, Nairobi, Kenya.

Zeven, A.C. and de Wet, J.M.J. (1983) *Dictionary of Cultivated Plants and their Regions of Diversity*, Centre for Agricultural Publishing and Documentation, Wageningen, The Netherlands.

Zhukovsky, P.M. (1975), *World Gene Pool of Plants for Breeding: Mega-gene-centres and Micro-gene-centres*, USSR Academy of Sciences, Leningrad, Soviet Union.

Zieg, R. et al. (1983) 'Selection of High Pyrethrin Producing Tissue Cultures', *Planta Medica*, Vol. 28, pp. 88–91.

Zirkle, C. (1949a) *Death of a Science in Russia: The Fate of Genetics as Described in Pravda and Elsewhere*, University of Pennsylvania Press, Philadelphia, USA.

——— (1949b) 'Plant Hybridization and Plant Breeding in Eighteenth-Century America', *Agricultural History,* Vol. 53, No. 1, pp. 25-38.

Zohary, D. (1970) 'Centres of Diversity and Centres of Origin', in Frankel, O.H. and Bennett, E. (eds) *Genetic Resources in Plants — Their Exploration and Conservation,* Blackwell Scientific Publications, Oxford, UK.

Zwanenberg, R. van and King, A. (1975) *An Economic History of Kenya and Uganda,* Macmillan Press, London, UK.

Index